501 Algebra Questions

OTHER TITLES OF INTEREST FROM
LEARNING EXPRESS

Algebra Success in 20 Minutes a Day
Geometry Success in 20 Minutes a Day
501 Geometry Questions
501 Math Word Problems

501 ALGEBRA QUESTIONS

3rd Edition

Mark A. McKibben, Ph.D.

LearningExpress®

NEW YORK

Published in the United States by LearningExpress, LLC, New York.

Library of Congress Cataloging-in-Publication Data:
501 algebra questions.—3rd ed.
p. cm.
ISBN 978-1-57685-898-1 (alk. paper)—ISBN 978-1-57685-898-7 (alk. paper)
1. Algebra—Problems, exercises, etc. I. Title: Five hundred one algebra questions.
II. Title: Five hundred and one algebra questions.
QA157.A15 2012
512—dc23
2012010087

Printed in the United States of America

9 8 7 6 5 4 3 2 1

Third Edition

ISBN 978-1-57685-898-1

For more information or to place an order, contact LearningExpress at:
2 Rector Street
26th Floor
New York, NY 10006

Or visit us at:
www.learningexpressllc.com

About the Author

Dr. Mark McKibben is currently a tenured associate professor of mathematics and computer science at Goucher College in Baltimore, Maryland. He earned his Ph.D. in mathematics from Ohio University in 1999, where his area of study was nonlinear analysis and differential equations. His dedication to undergraduate mathematics education prompted him to write textbooks and more than 20 supplements for courses on algebra, statistics, trigonometry, pre-calculus, and calculus. He is an active research mathematician who has published more than 25 original research articles, as well as a recent book entitled *Discovering Evolution Equations with Applications Volume 1: Deterministic Equations*, published by CRC Press/Chapman-Hall.

Contents

Introduction

This book is designed to provide you with review and practice for algebra success! It is not intended to teach common algebra topics. Instead, it provides 501 problems so you can flex your muscles and practice a variety of mathematical and algebraic skills. *501 Algebra Questions* is designed for many audiences. It's for anyone who has ever taken a course in algebra and needs to refresh and revive forgotten skills. It can be used to supplement current instruction in a math class or it can be used by teachers and tutors who need to reinforce student skills. If, at some point, you feel you need further explanation of some of the algebra topics highlighted in this book, you can find them in the LearningExpress publication *Algebra Success in 20 Minutes a Day*.

How to Use This Book

First, look at the table of contents to see the types of algebra topics covered in this book. The book is organized into 20 chapters with a variety of arithmetic, algebra, and word problems. The structure follows a common sequence of concepts introduced in basic algebra courses. You may want to follow the sequence, as each succeeding chapter builds on skills taught in previous chapters. But if your skills are just rusty, or if you are using this book to supplement topics you are currently learning, you may want to jump around from topic to topic.

Chapters are arranged using the same method. Each chapter has an introduction describing the mathematical concepts covered in the chapter. Second, there are helpful tips on how to practice the problems in each chapter. Last, you are presented with a variety of problems that generally range from easier to more difficult problems and their answer explanations. In many books, you are given one model problem and then asked to do many problems following that model. In this book, every problem has a complete step-by-step explanation for its solution. If you find yourself getting stuck solving a problem, you can look at the answer explanation and use it to help you understand the problem-solving process.

As you are solving problems, it is important to be as organized and sequential in your written steps as possible. The purpose of drills and practice is to make you proficient at solving problems. Like an athlete preparing for the next season or a musician warming up for a concert, you become skillful with practice. If, after completing all the problems in a section, you feel that you need more practice, redo the problems. It's not the answer that matters most—it's the process and the reasoning skills that you want to master.

You will probably want to have a calculator handy as you work through some of the sections. It's always a good idea to use it to check your calculations. If you have difficulty factoring numbers, the multiplication chart on the next page may help you. If you are unfamiliar with prime numbers, use the list on the next page so you won't waste time trying to factor numbers that can't be factored. And don't forget to keep lots of scrap paper on hand.

Make a Commitment

Success does not come without effort. Make the commitment to improve your algebra skills. Work for understanding. *Why* you do a math operation is as important as *how* you do it. If you truly want to be successful, make a commitment to spend the time you need to do a good job. You can do it! When you achieve algebra success, you have laid the foundation for future challenges and success. So sharpen your pencil and practice!

Multiplication Table

×	2	3	4	5	6	7	8	9	10	11	12
2	4	6	8	10	12	14	16	18	20	22	24
3	6	9	12	15	18	21	24	27	30	33	36
4	8	12	16	20	24	28	32	36	40	44	48
5	10	15	20	25	30	35	40	45	50	55	60
6	12	18	24	30	36	42	48	54	60	66	72
7	14	21	28	35	42	49	56	63	70	77	84
8	16	24	32	40	48	56	64	72	80	88	96
9	18	27	36	45	54	63	72	81	90	99	108
10	20	30	40	50	60	70	80	90	100	110	120
11	22	33	44	55	66	77	88	99	110	121	132
12	24	36	48	60	72	84	96	108	120	132	144

Commonly Used Prime Numbers

2	3	5	7	11	13	17	19	23	29
31	37	41	43	47	53	59	61	67	71
73	79	83	89	97	101	103	107	109	113
127	131	137	139	149	151	157	163	167	173
179	181	191	193	197	199	211	223	227	229
233	239	241	251	257	263	269	271	277	281
283	293	307	311	313	317	331	337	347	349
353	359	367	373	379	383	389	397	401	409
419	421	431	433	439	443	449	457	461	463
467	479	487	491	499	503	509	521	523	541
547	557	563	569	571	577	587	593	599	601
607	613	617	619	631	641	643	647	653	659
661	673	677	683	691	701	709	719	727	733
739	743	751	757	761	769	773	787	797	809
811	821	823	827	829	839	853	857	859	863
877	881	883	887	907	911	919	929	937	941
947	953	967	971	977	983	991	997	1,009	1,013

501 Algebra Questions

1 Working with Integers

For some people, it is helpful to try to simplify expressions containing signed numbers as much as possible. When you find signed numbers with addition and subtraction operations, you can simplify the task by changing all subtraction to addition. Subtracting a number is the same as adding its opposite. For example, subtracting a three is the same as adding a negative three. Or subtracting a negative 14 is the same as adding a positive 14. As you go through the step-by-step answer explanations, you will begin to see how this process of using only addition can help simplify your understanding of operations with signed numbers. As you begin to gain confidence, you may be able to eliminate some of the steps by doing them in your head and not having to write them down. After all, that's the point of practice! You work at the problems until the process becomes automatic. Then you own that process and you are ready to use it in other situations.

The **Tips for Working with Integers** section that follows gives you some simple rules to follow as you solve problems with integers. Refer to them each time you do a problem until you don't need to look at them. That is a sign of mastery.

You will also want to review the rules for Order of Operations with numerical expressions. You can use a memory device called a *mnemonic* to help you remember a set of instructions. Try remembering the acronym **PEMDAS** to help you remember the following process:

P do operations inside *Parentheses*
E evaluate terms with *Exponents*
M D do *Multiplication* and *Division* in order from left to right
A S *Add* and *Subtract* terms in order from left to right

Tips for Working with Integers

Addition

Signed numbers the same?: Find the SUM and use the same sign.
Signed numbers different?: Find the DIFFERENCE and use the sign of the larger number. (The larger number is the one whose value without a positive or negative sign is greatest.)

Addition is commutative. That is, you can add numbers in either order and the result is the same. As an example, **3 + 5 = 5 + 3**, or **$^-2 + {}^-1 = {}^-1 + {}^-2$**.

Subtraction

Change the operation sign to addition, change the sign of the number following the operation, then follow the rules for addition.

Multiplication/Division

Signs the same?: Multiply or divide and affix a **positive** sign to the product.
Signs different?: Multiply or divide and affix a **negative** sign to the product.

Multiplication is commutative. You can multiply terms in any order and the result will be the same. For example: **(2 • 5 • 7) = (2 • 7 • 5) = (5 • 2 • 7) = (5 • 7 • 2)** and so on.

For numbers 1–20, evaluate the given expression:

1. $27 + {}^-5$

2. $^-18 + {}^-20 - 16$

3. $^-15 - {}^-7$

4. $^-83 + 13$

5. $8 + {}^-4 - 12$

6. $38 \div {}^-2 + 9$

7. $^-25 \cdot {}^-3 + 15 \cdot {}^-5$

8. $24 \cdot {}^-8 + 2$

9. $^-5 \cdot {}^-9 \cdot {}^-2$

10. $^-2 \cdot {}^-3 \cdot {}^-4 \cdot {}^-5$

11. $^-10 - {}^-5 - 4 + {}^-1$

12. $(49 \div 7) - (48 \div {}^-4)$

13. $({}^-7 \cdot {}^-5) \div ({}^-4 - {}^-3)$

14. ${}^-(5 \cdot 3) + (12 \div {}^-4)$

15. $({}^-18 \div 2) - (6 \cdot {}^-3)$

16. $23 + (64 \div {}^-16)$

17. $2^3 - ({}^-4)^2$

18. $(3 - 5)^3 + (18 \div 6)^2$

19. $(3^2 + 6) \div ({}^-24 \div 8)$

20. $1 - [2 - 3(1 - 2^3)]$

For numbers 21–25, solve the given word problem.

21. A scuba diver descends 80 feet, rises 25 feet, descends 12 feet, and then rises 52 feet where he will do a safety stop for five minutes before surfacing. At what depth did he do his safety stop?

22. A digital thermometer records the daily high and low temperatures. The high for the day was $^+5°$ C. The low was $^-11°$ C. What was the difference between the day's high and low temperatures?

23. A checkbook balance sheet shows an initial balance for the month of \$300. During the month, checks were written in the amounts of \$25, \$82, \$213, and \$97. Deposits were also made into the account in the amounts of \$84 and \$116 during this time period. What was the balance at the end of the month?

24. A gambler begins playing a slot machine with \$10 in quarters in her coin bucket. She plays 15 quarters before winning a jackpot of 50 quarters. She then plays 20 more quarters in the same machine before walking away. How many quarters does she now have in her coin bucket?

25. A glider is towed to an altitude of 2,000 feet above the ground before being released by the tow plane. The glider loses 450 feet of altitude before finding an updraft that lifts it 1,750 feet. What is the glider's altitude now?

Answers

Numerical expressions in parentheses like this [ ] are operations performed on only part of the original expression. The operations performed within these symbols are intended to show how to evaluate the various terms that make up the entire expression.

Expressions with parentheses that look like this ( ) contain either numerical substitutions or expressions that are part of a numerical expression. Once a single number appears within these parentheses, the parentheses are no longer needed and need not be used the next time the entire expression is written.

When two pair of parentheses appear side by side like this ( )( ), it means that the expressions within are to be multiplied.

Sometimes parentheses appear within other parentheses in numerical or algebraic expressions. Regardless of what symbol is used, ( ), { }, or [ ], perform operations in the innermost parentheses first and work outward.

Underlined equations show the original algebraic equation equal to its simplified result.

1. The signs of the terms are different, so find the difference of the values. $[27 - 5 = 22]$
The sign of the larger term is positive, so the sign of the result is positive. $\underline{27 + {}^{-}5 = {}^{+}22}$

2. Change the subtraction sign to addition by changing the sign of the number that follows it. ${}^{-}18 + {}^{-}20 + ({}^{-}16)$
Since all the signs are negative, add the absolute value of the numbers. $[18 + 20 + 16 = 54]$
Since all the signs were negative, the result is negative. ${}^{-}18 + {}^{-}20 + {}^{-}16 = {}^{-}54$
The simplified result of the numeric expression is as follows: $\underline{{}^{-}18 + {}^{-}20 - 16 = {}^{-}54}$

3. Change the subtraction sign to addition by changing the sign of the number that follows it. ${}^{-}15 + 7$
Signs different? Subtract the absolute value of the numbers. $[15 - 7 = 8]$
Affix the sign of the larger term to the result. ${}^{-}15 + 7 = {}^{-}8$
The simplified expression is as follows: $\underline{{}^{-}15 - {}^{-}7 = {}^{-}8}$

4. The signs of the terms are different, so find the difference of the values. $[83 - 13 = 70]$
The sign of the larger term is negative, so the sign of the result is negative. $\underline{{}^{-}83 + 13 = {}^{-}70}$

5. Change the subtraction sign to addition by changing the sign of the number that follows it. $8 + {}^{-}4 + {}^{-}12$
With three terms, first group like terms and add. $8 + ({}^{-}4 + {}^{-}12)$

Signs the same? Add the value of the terms and give the result the same sign. $[(^{-}4 + ^{-}12) = ^{-}16]$
Substitute the result into the first expression. $8 + (^{-}16)$
Signs different? Subtract the value of the numbers. $[16 - 8 = 8]$
Give the result the sign of the larger term. $8 + (^{-}16) = ^{-}8$
The simplified result of the numeric expression is as follows: $\underline{8 + ^{-}4 - 12 = ^{-}8}$

6. First divide. Signs different? Divide and give the result the negative sign. $[(38 \div ^{-}2) = ^{-}19]$
Substitute the result into the expression. $(^{-}19) + 9$
Signs different? Subtract the value of the numbers. $[19 - 9 = 10]$
Give the result the sign of the term with the larger value. $(^{-}19) + 9 = ^{-}10$
The simplified result of the numeric expression is as follows: $\underline{38 \div ^{-}2 + 9 = ^{-}10}$

7. First perform the two separate multiplications.
Signs the same? Multiply the terms and give the result a positive sign. $[^{-}25 \cdot ^{-}3 = ^{+}75]$
Signs different? Multiply the terms and give the result a negative sign. $[15 \cdot ^{-}5 = ^{-}75]$
Now substitute the results into the original expression. $(^{+}75) + (^{-}75)$
Signs different? Subtract the value of the numbers. $[75 - 75 = 0]$
The simplified result of the numeric expression is as follows: $\underline{^{-}25 \cdot ^{-}3 + 15 \cdot ^{-}5 = 0}$

8. Group the terms being multiplied and evaluate. $(24 \cdot ^{-}8) + 2$
Signs different? Multiply the terms and give the result a negative sign. $[24 \cdot ^{-}8 = ^{-}192]$
Substitute the result into the original expression. $(^{-}192) + 2$
Signs different? Subtract the value of the terms. $[192 - 2 = 190]$
Give the result the sign of the term with the larger value. $(^{-}192) + 2 = ^{-}190$
The simplified result of the numeric expression is as follows: $\underline{24 \cdot ^{-}8 + 2 = ^{-}190}$

9. Because all the operators are multiplication, you could multiply the numbers in any order and the result would be the same. Let's group the first two terms. $(^{-}5 \cdot ^{-}9) \cdot ^{-}2$
Signs the same? Multiply the terms and give the result a positive sign. $[^{-}5 \cdot ^{-}9 = ^{+}45]$

Now substitute the result into the original expression.	$^{+}45 \cdot {}^{-}2$
Signs different? Multiply the terms and give the result a negative sign.	$^{+}45 \cdot {}^{-}2 = {}^{-}90$
The simplified result of the numeric expression is as follows:	$^{-}5 \cdot {}^{-}9 \cdot {}^{-}2 = {}^{-}90$

10.

Because all the operators are multiplicative, you could group any two terms and the result would be the same. Let's group the first two terms and the last to terms.	$({}^{-}2 \cdot {}^{-}3) \cdot ({}^{-}4 \cdot {}^{-}5)$
Signs the same? Multiply the terms and give the result a positive sign.	$[({}^{-}2 \cdot {}^{-}3) = {}^{+}6]$ $[({}^{-}4 \cdot {}^{-}5) = {}^{+}20]$
Substitute.	$({}^{+}6) \cdot ({}^{+}20)$
Signs the same? Multiply the terms and give the result a positive sign.	$({}^{+}6) \cdot ({}^{+}20) = {}^{+}120$
The simplified result of the numeric expression is as follows:	$^{-}2 \cdot {}^{-}3 \cdot {}^{-}4 \cdot {}^{-}5 = {}^{+}120$

11.

Proceed left to right with the addition and subtraction as they rise.	
Change the subtraction sign to addition by changing the sign of the number that follows it.	$[{}^{-}10 - {}^{-}5) = {}^{-}10 + 5]$
Signs different? Subtract the absolute value of the numbers.	$[10 - 5 = 5]$
Affix the sign of the larger term to the result.	$[{}^{-}10 + 5 = {}^{-}5]$
Substitute.	$^{-}5 - 4 + {}^{-}1$
Change the subtraction sign to addition by changing the sign of the number that follows it.	$[{}^{-}5 - 4 = {}^{-}5 + {}^{-}4]$
Signs the same? Add the absolute value of the terms.	$[5 + 4 = 9]$
Affix the sign of the terms to the result.	$[{}^{-}5 + {}^{-}4 = {}^{-}9]$
Substitute.	$^{-}9 + {}^{-}1$
Signs the same? Add the absolute value of the terms.	$[9 + 1 = 10]$
Affix the sign of the terms to the result.	$^{-}9 + {}^{-}1 = {}^{-}10$
The simplified result of the numeric expression is as follows:	$^{-}10 - {}^{-}5 - 4 + {}^{-}1 = {}^{-}10$

12.

First evaluate the expressions within each set of parentheses.	$[49 \div 7 = 7]$
Signs different? Divide and give the result a negative sign.	$[48 \div {}^{-}4 = {}^{-}12]$
Substitute into the original expression.	$(7) - ({}^{-}12)$
Change the subtraction sign to addition by changing the sign of the number that follows it.	$7 + {}^{+}12$
Signs the same? Add the value of the terms and give the result the same sign.	$7 + {}^{+}12 = {}^{+}19$

The simplified result of the numeric expression is as follows: $(49 \div 7) - (48 \div {}^-4) = {}^+19$

13. First, simplify the results in each set of parentheses.
For the first set of parentheses, signs the same? Multiply the terms and give the result a positive sign: $[({}^-7 \cdot {}^-5) = {}^+35]$
For the second set of parentheses, change the subtraction sign to addition by changing the sign of the number that follows it: $[({}^-4 - {}^-3) = {}^-4 + 3]$
Signs different? Subtract the absolute value of the number. $[4 - 3 = 1]$
Affix the sign of the larger term. $[{}^-4 + 3 = {}^-1]$
The original expression now simplifies to: $35 \div {}^-1 = {}^-35$

14. First evaluate the expressions within each set of parentheses. $[5 \cdot 3 = 15]$
Signs different? Divide and give the result a negative sign. $[12 \div {}^-4 = {}^-3]$
Substitute the values into the original expression. ${}^-(15) + ({}^-3)$
Signs the same? Add the value of the terms and give the result the same sign. $[15 + 3 = 18]$
${}^-(15) + ({}^-3) = {}^-18$

The simplified result of the numeric expression is as follows: ${}^-(5 \cdot 3) + (12 \div {}^-4) = {}^-18$

15. First evaluate the expressions within each set of parentheses. $[({}^-18 \div 2)]$
Signs different? Divide the value of the terms and give the result a negative sign. $[18 \div 2 = 9]$
$[({}^-18 \div 2 = {}^-9)]$

Signs different? Multiply the term values and give the result a negative sign. $(6 \cdot {}^-3)$
$[6 \cdot 3 = 18]$
$(6 \cdot {}^-3) = {}^-18$

Substitute the values into the original expression. $({}^-9) - ({}^-18)$
Change subtraction to addition and change the sign of the term that follows. $({}^-9) + ({}^+18)$
Signs different? Subtract the value of the numbers and give the result the sign of the higher value number. $[18 - 9 = 9]$
$({}^-9) + ({}^+18) = {}^+9$

The simplified result of the numeric expression is as follows: $({}^-18 \div 2) - (6 \cdot {}^-3) = {}^+9$

16.

Evaluate the expression within the parentheses.	$(64 \div {}^{-}16)$
Signs different? Divide and give the result a negative sign.	$[64 \div 16 = 4]$
	$(64 \div {}^{-}16 = {}^{-}4)$
Substitute the value into the original expression.	$23 + ({}^{-}4)$
Signs different? Subtract the value of the numbers and give the result the sign of the higher value number.	$[23 - 4 = 19]$
	$23 + ({}^{-}4) = {}^{+}19$
The simplified result of the numeric expression is as follows:	$\underline{23 + (64 \div {}^{-}16) = {}^{+}19}$

17.

The order of operations tells us to evaluate the terms with exponents first.	$[2^3 = 2 \cdot 2 \cdot 2 = 8]$
	$[({}^{-}4)^2 = ({}^{-}4) \cdot ({}^{-}4)]$
Signs the same? Multiply the terms and give the result a positive sign.	$[4 \cdot 4 = 16]$
	$[({}^{-}4)^2 = {}^{+}16]$
Substitute the values of terms with exponents into the original expression.	$2^3 - ({}^{-}4)^2 = (8) - ({}^{+}16)$
Change subtraction to addition and change the sign of the term that follows.	$8 + {}^{-}16$
Signs different? Subtract the value of the numbers and give the result the sign of the higher value number.	$[16 - 8 = 8]$
	$8 + {}^{-}16 = {}^{-}8$
The simplified result of the numeric expression is as follows:	$\underline{2^3 - ({}^{-}4)^2 = {}^{-}8}$

18.

First evaluate the expressions within each set of parentheses.	$[3 - 5]$
Change subtraction to addition and change the sign of the term that follows.	$[3 + ({}^{-}5)]$
Signs different? Subtract the value of the numbers and give the result the sign of the higher value number.	$[5 - 3 = 2]$
	$[3 - 5 = {}^{-}2]$
	$[18 \div 6 = 3]$
Substitute the values of the expressions in parentheses into the original expression.	$({}^{-}2)^3 + (3)^2$
Evaluate the terms with exponents.	$[({}^{-}2)^3 = {}^{-}2 \cdot {}^{-}2 \cdot {}^{-}2]$
	$[({}^{-}2 \cdot {}^{-}2) \cdot {}^{-}2 = ({}^{+}4) \cdot {}^{-}2]$
Signs different? Multiply the value of the terms and give the result a negative sign.	$[({}^{+}4) \cdot {}^{-}2 = {}^{-}8]$
	$[(3)^2 = 3 \cdot 3 = 9]$
Substitute the values into the expression.	$({}^{-}2)^3 + (3)^2 = {}^{-}8 + 9$

Signs different? Subtract the value of the numbers and give the result the sign of the higher value number. $9 - 8 = {}^{+}1$

The simplified result of the numeric expression is as follows: $\underline{(3 - 5)^3 + (18 \div 6)^2 = {}^{+}1}$

19. First evaluate the expressions within each set of parentheses. $[3^2 + 6 = (9) + 6 = 15]$

Signs different? Divide and give the result the negative sign. $[{}^{-}24 \div 8 = {}^{-}3]$

Substitute values into the original expression. $(15) \div ({}^{-}3)$

Signs different? Divide the value of the terms and give the result a negative sign. $[15 \div 3 = 5]$

$(15) \div ({}^{-}3) = {}^{-}5$

The simplified result of the numeric expression is as follows: $\underline{(3^2 + 6) \div ({}^{-}24 \div 8) = {}^{-}5}$

20. Evaluate the innermost parentheses first.

First, compute the power. $[(1 - 2^3) = (1 - 8)]$

Signs different? Subtract the absolute value of the numbers. $[8 - 1 = 7]$

Affix the sign of the larger term to the result. $[(1 - 8) = {}^{-}7]$

Substitute. $1 - [2 - 3({}^{-}7)]$

Now, evaluate the brackets, beginning with the multiplication.

Signs different? Multiply the terms and give the result a negative sign: $[3({}^{-}7) = {}^{-}21]$

Substitute: $1 - [2 - ({}^{-}21)]$

Now, subtract the terms within brackets. Change the subtraction sign to addition by changing the sign of the number that follows it. $[2 - ({}^{-}21) = 2 + 21]$

Signs the same, and both positive? Add the numbers and affix a positive sign. $[2 + 21 = 23]$

Substitute. $1 - [2 - ({}^{-}21)] = 1 - 23$

Finally, change the subtraction sign to addition by changing the sign of the number that follows it. $[1 - 23 = 1 + {}^{-}23]$

Signs different? Subtract the absolute value of the numbers. $[23 - 1 = 22]$

Affix the sign of the larger term to the result. $[1 - 23 = {}^{-}22]$

The original expression now simplifies to: $\underline{1 - [2 - 3(1 - 2^3)] = {}^{-}22}$

21. If you think of distance above sea level as a positive number, then going below sea level is represented as a negative number. Going up is in the positive direction, while going down is in the negative direction. Give all the descending distances a negative sign and the ascending distances a positive sign.

The resulting numerical expression would be as follows: $^{-}80 + ^{+}25 + ^{-}12 + ^{+}52$

Because addition is commutative, you can associate like-signed numbers. $(^{-}80 + ^{-}12) + (^{+}25 + ^{+}52)$

Evaluate the numerical expression in each parentheses. $[^{-}80 + ^{-}12 = ^{-}92]$ $[^{+}25 + ^{+}52 = ^{+}77]$

Substitute the values into the numerical expression. $(^{-}92) + (^{+}77)$

Signs different? Subtract the value of the numbers and give the result the sign of the higher value number. $[92 - 77 = 15]$

So, $^{-}80 + ^{+}25 + ^{-}12 + ^{+}52 = ^{-}52$

The diver took his rest stop at $^{-}15$ feet.

22. You could simply figure that $^{+}5°$ C is 5° above zero and $^{-}11°$ C is 11° below. So the difference is the total of 5° + 11° = 16°.

Or you could find the difference between $^{+}5°$ and $^{-}11°$.

That would be represented by the following equation. $^{+}5° - ^{-}11° = ^{+}5° + ^{+}11° = ^{+}16°$

23. You can consider that balances and deposits are positive signed numbers, while checks are deductions, represented by negative signed numbers.

An expression to represent the activity during the month would be as follows: $300 + ^{-}25 + ^{-}82 + ^{-}213 + ^{-}97 + ^{+}84 + ^{+}116$

Because addition is commutative, you can associate like signed numbers. $(300 + ^{+}84 + ^{+}116) + (^{-}25 + ^{-}82 + ^{-}213 + ^{-}97)$

Evaluate the numbers within each parentheses. $[300 + ^{+}84 + ^{+}116 = ^{+}500]$ $[(^{-}25 + ^{-}82 + ^{-}213 + ^{-}97) = ^{-}417]$

Substitute the values into the revised expression. $(^{+}500) + (^{-}417) = ^{+}83$

The balance at the end of the month would be \$83.

24. You first figure out how many quarters she starts with. Four quarters per dollar gives you $4 \cdot 10 = 40$ quarters. You can write an expression that represents the quarters in the bucket and the quarters added and subtracted. In chronological order, the expression would be as follows: $40 - 15 + 50 - 20$

Change all operation signs to addition and the sign of the number that follows. $40 + ^{-}15 + 50 + ^{-}20$

Because addition is commutative, you can associate like-signed numbers. $(40 + 50) + (^{-}15 + ^{-}20)$

Use the rules for adding integers with like signs.	$[40 + 50 = 90]$ $[^{-}15 + {}^{-}20 = {}^{-}35]$
Substitute into the revised expression.	$(90) + (^{-}35)$
Signs different? Subtract the value of the numbers and give the result the sign of the higher value number.	$[90 - 35 = 55]$
The simplified result of the numeric expression is as follows:	$40 + {}^{-}15 + 50 + {}^{-}20 = 55$

<u>She has 55 quarters in her coin bucket.</u>

25. As in problem 21, ascending is a positive number while descending is a negative number. You can assume ground level is the zero point.

An expression that represents the problem is as follows:	$^{+}2{,}000 + {}^{-}450 + {}^{+}1{,}750$
Because addition is commutative, you can associate like-signed numbers.	$(^{+}2{,}000 + {}^{+}1{,}750) + {}^{-}450$
Evaluate the expression in the parentheses.	$[^{+}2{,}000 + {}^{+}1{,}750 = {}^{+}3{,}750]$
Substitute into the revised equation.	$(^{+}3{,}750) + {}^{-}450$
Signs different? Subtract the value of the numbers and give the result the sign of the higher value number.	$[3{,}750 - 450 = 3{,}300]$
The simplified result of the numeric expression is as follows:	$(^{+}3{,}750) + {}^{-}450 = {}^{+}3{,}300$

<u>The glider's final altitude is 3,330 ft.</u>

2

Working with Algebraic Expressions

This chapter contains 25 algebraic expressions; each can contain up to five variables. Remember that a variable is just a letter that represents a number in a mathematical expression. When given numerical values for the variables, you can turn an algebraic expression into a numerical one.

As you work through the problems in this chapter, you are to substitute the assigned values for the variables into the expression and evaluate the resulting expression. You will be evaluating expressions very much like the previous numerical expressions. The answer section contains complete explanations of how to go about evaluating the expressions. Work on developing a similar style throughout, and you will have one sure way of solving these kinds of problems. As you become more familiar and comfortable with the look and feel of these expressions, you will begin to find your own shortcuts.

Read through the **Tips for Working with Algebraic Expressions** before you begin to solve the problems in this section.

Tips for Working with Algebraic Expressions

- Substitute assigned values for the variables into the expression.
- Use PEMDAS to perform operations in the proper order.
- Recall and use the **Tips for Working with Integers** from Chapter 1.

- Evaluate the following algebraic expressions when

$a = 3$
$b = {}^{-}5$
$x = 6$
$y = \frac{1}{2}$
$z = {}^{-}8$

26. $4a + z$

27. $3x \div z$

28. $2ax - z$

29. ${}^{-}3(xy - 2ab)$

30. $4b^2 - az$

31. $7x \div 2yz$

32. $bx + z \div y$

33. $({}^{-}z + 2b) \cdot (a - x)$

34. $a(b + z)^2$

35. $2(a^2 + 2y) \div b$

36. $a^3 + 24y - 3b$

37. ${}^{-}2x - b + az$

38. $5z^2 - 2z + 2$

39. $(b + xy)^3 - (b + xy)^2$

40. $7x + \frac{12}{x} - z$

41. $2b^2 \div y$

42. $bx(z + 3)$

43. $6y(z \div y) + 3ab$

44. $2bx \div (z - b)$

45. $(\frac{1}{4}yz)^5$

46. $y\{[\frac{x}{2} - 3] - 4a\}$

47. $10b^3 - 4b^2$

48. $-(b + x + y)^2$

49. $z^2 - 4a^2y$

50. $3x^2b[5a - 3b]$

Answers

Numerical expressions in parentheses like this [ ] are operations performed on only part of the original expression. The operations performed within these symbols are intended to show how to evaluate the various terms that make up the entire expression.

Expressions with parentheses that look like this ( ) contain either numerical substitutions or expressions that are part of a numerical expression. Once a single number appears within these parentheses, the parentheses are no longer needed and need not be used the next time the entire expression is written.

When two pair of parentheses appear side by side like this ( )( ), it means that the expressions within are to be multiplied.

Sometimes parentheses appear within other parentheses in numerical or algebraic expressions. Regardless of what symbol is used, ( ), { }, or [ ], perform operations in the innermost parentheses first and work outward.

<u>Underlined</u> equations show the original algebraic expression equal to its numerical value.

$a = 3$
$b = {}^-5$
$x = 6$
$y = \frac{1}{2}$
$z = {}^-8$

26. Substitute the values for the variables into the expression. $4(3) + ({}^-8)$
Order of operations tells you to multiply first. $[4(3) = 12]$
Substitute. $(12) + ({}^-8)$
Signs different? Subtract the value of the numbers. $12 - 8 = 4$
Give the result the sign of the larger value. (No sign means $^+$) $^+4$
The value of the expression is as follows: <u>$4a + z = {}^+4$</u>

27. Substitute the values for the variables into the expression. $3(6) \div ({}^-8)$
PEMDAS: Multiply the first term. $[3(6) = 18]$
Substitute. $(18) \div ({}^-8)$
Signs different? Divide and give the result the negative sign. $[18 \div 8 = 2\frac{2}{8} = 2\frac{1}{4}]$
$({}^-2\frac{1}{4})$
The value of the expression is as follows: <u>$3x \div z = {}^-2\frac{1}{4}$ or $^-2.25$</u>

28. Substitute the values for the variables into the expression. $2(3)(6) - ({}^-8)$
Multiply the factors of the first term. $[2(3)(6) = 36]$
Substitute. $(36) - ({}^-8)$
Change the operator to addition and the sign of the number that follows. $(36) + ({}^+8)$
Signs the same? Add the value of the terms and give the result the same sign. $[36 + 8 = 44]$
$({}^+44)$

The simplified value of the expression is as follows: $2ax - z = {}^{+}44$

29. Substitute the values for the variables into the expression. $^{-}3((6))(\frac{1}{2}) - 2(3)(^{-}5))$

Order of operations tells you to simplify the expression within the parentheses first.

Multiply the first two terms within the parentheses. $[(6)(\frac{1}{2}) = \frac{6}{1} \cdot \frac{1}{2} = \frac{6 \cdot 1}{1 \cdot 2} = 3]$

Multiply the last three terms. We can group the terms any way we like, so let's group the first two together. Then, multiply that by the third term. $[2(3) = 6]$ $[6(^{-}5) = {}^{-}30]$

Substitute. $^{-}3(3 - (^{-}30))$

Next, perform the subtraction within the parentheses.

Change the operator to addition and the sign of the number that follows. $[3 - (^{-}30) = 3 + 30]$

Signs the same? Add the value of the terms and give the result the same sign. $[3 + 30 = 33]$

Substitute. $^{-}3(33)$

Finally, multiply the terms. Signs different? Multiplythe numbers and give the result a negative sign. $[^{-}3(33) = {}^{-}99]$

The original expression now simplifies to: $^{-}3((6))(\frac{1}{2}) - 2(3)(^{-}5)) = {}^{-}99$

30. Substitute the values for the variables into the expression. $4(^{-}5)^2 - (3)(^{-}8)$

PEMDAS: Evaluate the term with the exponent. $[(^{-}5)^2 = (^{-}5) \cdot (^{-}5)]$

Signs the same? Multiply the terms and give the result a positive sign. $[5 \cdot 5 = 25 = {}^{+}25]$

Substitute. $[4(^{+}25) = 100]$

Now evaluate the other term. $[(3)(^{-}8)]$

Signs different? Multiply and give the result a negative sign. $[3 \cdot {}^{-}8 = {}^{-}24]$

Substitute the equivalent values into the original expression. $(100) - (^{-}24)$

Change the operator to addition and the sign of the number that follows. $100 + {}^{+}24 = 124$

The simplified value of the expression is as follows: $4b^2 - az = 124$

31. Substitute the values for the variables into the expression. $7(6) \div 2(\frac{1}{2})(^{-}8)$

PEMDAS: Multiply the terms in the expression. $[7 \cdot 6 = 42]$

$[\{2(\frac{1}{2})\}(^{-}8) = (1)(^{-}8) = {}^{-}8]$

Substitute the equivalent values into the original expression. $(42) \div (^{-}8)$

Signs different? Divide and give the result a negative sign. $[42 \div 8 = 5.25]$
$(42) \div (^-8) = {}^-5.25$

The simplified value of the expression is as follows: $\underline{7x \div 2yz = {}^-5.25}$

32. Substitute the values for the variables into the expression. $(^-5)(6) + (^-8) \div (\frac{1}{2})$

Group terms using order of operations. $(^-5)(6) + \{(^-8) \div (\frac{1}{2})\}$

PEMDAS: Multiply or divide the terms in the expression from left to right as they arise. $[(^-5)(6)]$

Signs different? Multiply and give the result a negative sign. $[5 \cdot 6 = 30]$
$[(^-5)(6) = {}^-30]$

Consider the second term. $[(^-8) \div (\frac{1}{2})]$

Signs different? Divide and give the result a negative sign. $[8 \div \frac{1}{2}]$

To divide by a fraction, you multiply by its reciprocal. $[8 \div \frac{1}{2} = 8 \cdot \frac{2}{1} = 8 \cdot 2 = 16]$
$[(^-8) \div (\frac{1}{2}) = {}^-16]$

Substitute the equivalent values into the original expression. $(^-30) + (^-16)$

Signs the same? Add the value of the terms and give the result the same sign. $[30 + 16 = 46]$
$(^-30) + (^-16) = {}^-46$

The simplified value of the expression is as follows: $\underline{bx + z \div y = {}^-46}$

33. Substitute the values for the variables into the expression. $(-(^-8) + 2(^-5)) \cdot (3 - 6)$

Order of operations tells you first to simplify the expressions enclosed within the parentheses. Begin with the first set of parentheses.

Simplify the first term. $[-(^-8) = {}^+8]$

Multiply the last two terms. Signs different? Multiply the numbers and give the result a negative sign. $[2(^-5) = {}^-10]$

Substitute. $(8 - 10) \cdot (3 - 6)$

Now, compute the difference in the first parentheses.

Change the operator to addition and the sign of the number that follows it. $[8 - 10 = 8 + {}^-10]$

Subtract the numbers and give the result the sign of the larger term. $[8 + {}^-10 = {}^-2]$

Substitute. $(^-2) \cdot (3 - 6)$

Compute the difference in the second set of parentheses.

Change the operator to addition and the sign of the number that follows it. $[3 - 6 = 3 + {}^-6]$

Subtract the numbers and give the result the sign of the larger term. $[3 + {}^-6 = {}^-3]$

Substitute. $(^-2) \cdot (^-3)$

Finally, compute the product. Signs the same? Give the result a positive sign. $[(^-2) \cdot (^-3) = 6]$

The original expression now simplifies to: $\underline{(-(^-8) + 2(^-5)) \cdot (3 - 6) = 6}$

34. Substitute the values for the variables into the expression. $3((^-5) + (^-8))^2$

PEMDAS: You must add the terms inside the parentheses first. $[(^-5) + (^-8)]$

Signs the same? Add the value of the terms and give the result the same sign. $[5 + 8 = 13]$

$[(^-5) + (^-8) = {}^-13]$

Substitute into the original expression. $3(^-13)^2$

Next you evaluate the term with the exponent. $[(^-13)^2 = {}^-13 \cdot {}^-13]$

Signs the same? Multiply the terms and give the result a positive sign. $[13 \cdot 13 = {}^+169]$

Substitute the equivalent values into the original expression. $3(169) = 507$

The simplified value of the expression is as follows: $\underline{a(b + z)^2 = 507}$

35. Substitute the values for the variables into the expression. $2\mathbf{(}(3)^2 + 2(\frac{1}{2})\mathbf{)} \div (^-5)$

First, evaluate the term inside the bold parentheses. $[(3)^2 + 2(\frac{1}{2})]$

The first term has an exponent. Evaluate it. $[(3)^2 = 3 \cdot 3 = 9]$

Evaluate the second term. $[2(\frac{1}{2}) = 1]$

$[(3)^2 + 2(\frac{1}{2}) = 9 + 1 = 10]$

Substitute into the original numerical expression. $2(10) \div (^-5)$

Evaluate the first term. $[2(10) = 20]$

Substitute into the numerical expression. $(20) \div (^-5)$

Signs different? Divide and give the result a negative sign. $[20 \div 5 = 4]$

$(20) \div (^-5) = {}^-4$

The simplified value of the expression is as follows: $\underline{2(a^2 + 2y) \div b = {}^-4}$

36. Substitute the values for the variables into the expression. $(3)^3 + 24(\frac{1}{2}) - 3(^-5)$

Evaluate the term with the exponent. $[(3)^3 = 3 \cdot 3 \cdot 3 = 27]$

Evaluate the second term. $[24(\frac{1}{2}) = 12]$

Evaluate the third term. $[3(^-5)]$

Signs different? Multiply and give the result a negative sign. $[3 \cdot 5 = 15]$

$[3(^-5) = {^-15}]$

Substitute the equivalent values into the original expression. $(27) + (12) - (^-15)$

Change the subtraction to addition and the sign of the number that follows. $(27) + (12) + (^+15)$

Signs the same? Add the value of the terms and give the result the same sign. $27 + 12 + 15 = {^+54}$

The simplified value of the expression is as follows: $\underline{a^3 + 24y - 3b = 54}$

37. Substitute the values for the variables into the expression. $^-2(6) - (^-5) + (3)(^-8)$

Evaluate the first and last terms. Positive times negative results in a negative. $[^-2(6) = {^-2} \cdot 6 = {^-12}]$

$[(3)(^-8) = 3 \cdot {^-8} = {^-24}]$

Substitute the equivalent values into the original expression. $(^-12) - (^-5) + (^-24)$

Change the subtraction to addition and the sign of the number that follows. $^-12 + (^+5) + {^-24}$

Commutative property of addition allows grouping of like signs. $(^-12 + {^-24}) + (^+5)$

Signs the same? Add the value of the terms and give the result the same sign. $[^-12 + {^-24} = {^-36}]$

Substitute. $(^-36) + (^+5)$

Signs different? Subtract and give the result the sign of the higher value number. $[36 - 5 = 31]$

$(^-36) + (^+5) = {^-31}$

The simplified value of the expression is as follows: $\underline{^-2x - b + az = {^-31}}$

38. Substitute the values for the variables into the expression. $5(^-8)^2 - 2(^-8) + 2$

PEMDAS: Evaluate the term with the exponent first. $[(^-8)^2 = (^-8)(^-8) = {^+64}]$

Substitute the value into the numerical expression. $5(^+64) - 2(^-8) + 2$

PEMDAS: Evaluate terms with multiplication next. $[5(^+64) = 320]$

$[2(^-8) = {}^-16]$

Substitute the values into the numerical expression. $320 - (^-16) + 2$

Change the subtraction to addition and the sign of the number that follows. $320 + (^+16) + 2$

Add terms from left to right. All term signs are positive, a result of addition $^+$. $320 + 16 + 2 = 338$

The simplified value of the expression is as follows: $\underline{5z^2\ 2z + 2 = 338}$

39. Substitute the values for the variables into the expression. $(^-5 + (6)(\frac{1}{2}))^3 - (^-5 + (6)(\frac{1}{2}))^2$

Order of operations tells you first to simplify the expressions enclosed within the parentheses. **NOTE!** The expressions are the same in both parentheses, so we will simplify them together rather than doing the work twice.

Multiply the last two terms. Signs the same? Give the result a positive sign. $[(6)(\frac{1}{2}) = \frac{6}{1} \cdot \frac{1}{2} = \frac{6 \cdot 1}{1 \cdot 2} = 3]$

Substitute. $(^-5 + 3)^3 - (^-5 + 3)^2$

Next, compute the sum enclosed in the parentheses. Signs different? Subtract the values and give the result the sign of the larger term. $[5 - 3 = 2]$ $[^-5 + 3 = {}^-2]$

Substitute. $(^-2)^3 - (^-2)^2$

Next, compute the two powers separately. $[(^-2)^3 = (^-2)(^-2)(^-2) = (4)(^-2) = {}^-8]$ $\underline{[(^-2)^2 = (^-2)(^-2) = 4]}$

Substitute. $^-8 - 4$

Finally, compute the difference. Change the operator to addition and the sign of the number that follows it. $[^-8 - 4 = {}^-8 + {}^-4]$

Signs the same? Add the terms and give the result the same sign. $[^-8 + {}^-4 = {}^-12]$

The original expression now simplifies to: $\underline{(^-5 + (6)(\frac{1}{2}))^3 - (^-5 + (6)(\frac{1}{2}))^2 = {}^-12}$

40. Substitute the values for the variables into the expression. $7(6) + \frac{12}{(6)} - (^-8)$

Evaluate the first term. $[7(6) = 7 \cdot 6 = 42]$

Evaluate the second term. $[\frac{12}{(6)} = 12 \div 6 = 2]$

Substitute the values into the original numerical expression. $(42) + (2) - {}^-8$

Change the subtraction to addition and the sign of the number that follows. $42 + 2 + {}^+8$

Add terms from left to right. $42 + 2 + {}^{+}8 = 52$
The simplified value of the expression is as follows: $\underline{7x + \frac{12}{x} - z = 52}$

41. Substitute the values for the variables in the expression. $2({}^{-}5)^2 \div \frac{1}{2}$
First, evaluate the term with the exponent. $[({}^{-}5)^2 = ({}^{-}5)({}^{-}5) = 25]$
Substitute. $(2)(25) \div \frac{1}{2}$
Order of operations tells us to multiply and divide as it arises from left to right.
So, multiply the first two terms. $[(2)(25) = 50]$
Substitute the values into the original numerical expression. $({}^{+}50) \div \frac{1}{2}$
Change division to multiplication and change the value to its reciprocal. $({}^{+}50) \cdot 2 = 100$
The simplified value of the expression is as follows: $\underline{2b^2 \div y = 100}$

42. Substitute the values for the variables into the expression. $({}^{-}5)(6)[({}^{-}8) + 3]$
First, evaluate the expression inside the brackets. $[({}^{-}8) + 3]$
Signs different? Subtract and give the result the sign of the higher value number. $[8 - 3 = 5]$
$[({}^{-}8) + 3 = {}^{-}5]$
Substitute the result into the numerical expression. $({}^{-}5)(6)({}^{-}5)$
Multiply from left to right. Negative times positive equals negative. $[{}^{-}5 \cdot 6 = {}^{-}30]$
Signs the same? Multiply and give the result a positive sign. $({}^{-}30) \cdot {}^{-}5 = {}^{+}150$
The simplified value of the expression is as follows: $\underline{bx(z + 3) = 150}$

43. Substitute the values for the variables into the expression. $6(\frac{1}{2})[{}^{-}8 \div \frac{1}{2}] + 3(3)({}^{-}5)$
First evaluate the expression inside the brackets.
Division by a fraction is the same as multiplication by its reciprocal. $[{}^{-}8 \div \frac{1}{2} = {}^{-}8 \cdot \frac{2}{1} = {}^{-}16]$
Substitute the result into the numerical expression. $6(\frac{1}{2})({}^{-}16) + 3(3)({}^{-}5)$
Evaluate the first term in the expression. $[6(\frac{1}{2})({}^{-}16) = 6 \cdot \frac{1}{2} \cdot {}^{-}16]$
$[3 \cdot {}^{-}16 = {}^{-}48]$
Evaluate the second term in the expression. $[3(3)({}^{-}5) = 3 \cdot 3 \cdot {}^{-}5]$
$[9 \cdot {}^{-}5 = {}^{-}45]$
Substitute the result into the numerical expression. $({}^{-}48) + ({}^{-}45)$
Signs the same? Add the value of the terms and give the result the same sign. ${}^{-}48 + {}^{-}45 = {}^{-}93$
The simplified value of the expression is as follows: $\underline{6y(z \div y) + 3ab = {}^{-}93}$

44. Substitute the values for the variables into the expression. $2(^-5)(6) \div [(^-8) - (^-5)]$

First evaluate the expression inside the brackets. $[(^-8 - ^-5)]$

Change the subtraction to addition and the sign of the number that follows. $[^-8 + 5]$

Signs different? Subtract and give the result the sign of the higher value number. $[8 - 5 = 3]$
$[^-8 + ^+5 = ^-3]$

Substitute the results into the numerical expression. $2(^-5)(6) \div (^-3)$

Multiply from left to right. $[2 \cdot ^-5 \cdot 6 = ^-60]$

Substitute the result into the numerical expression. $(^-60) \div ^-3$

Signs the same? Divide and give the result a positive sign. $[60 \div 3 = 20]$
$(^-60) \div (^-3) = 20$

The simplified value of the expression is as follows: $\underline{2bx \div (z - b) = 20}$

45. Substitute the values for the variables into the expression. $[(\frac{1}{4})(\frac{1}{2})(^-8)]^5$

Order of operations tells you first to simplify the expressions enclosed within the brackets.

Multiply the terms together, from left to right as they arise. $[(\frac{1}{4})(\frac{1}{2}) = \frac{1 \cdot 1}{4 \cdot 2} = \frac{1}{8}]$
$[\frac{1}{8}(^-8) = ^-1]$

Substitute. $[^-1]^5$

Finally, compute the power. $[^-1]^5 = (^-1)(^-1)(^-1)(^-1)(^-1) = ^-1$

The original expression now simplifies to: $\underline{[(\frac{1}{4})(\frac{1}{2})(^-8)]^5 = ^-1}$

46. Substitute the values for the variables into the expression. $(\frac{1}{2})\{[\frac{(6)}{2} - 3] - 4(3)\}$

Evaluate the expression in the innermost square brackets. $[[\frac{(6)}{2} - 3] = \frac{6}{2} - 3]$

PEMDAS: Division before subtraction. $[\frac{6}{2} - 3 = 3 - 3 = 0]$

Substitute the result into the numerical expression. $(\frac{1}{2})\{(0) - 4(3)\}$

Evaluate the expression inside the curly brackets. $[\{0 - 4(3)\} = 0 - 4 \cdot 3]$

PEMDAS: Multiply before subtraction. $[0 - 4 \cdot 3 = 0 - 12]$

Change subtraction to addition and the sign of the term that follows. $[0 - 12 = 0 + ^-12 = ^-12]$

Substitute the result into the numerical expression. $(\frac{1}{2})\{^-12\} = \frac{1}{2} \cdot ^-12$

Signs different? Multiply numbers and give the result a negative sign. $[\frac{1}{2} \cdot 12 = 6]$

$\frac{1}{2} \cdot {}^-12 = {}^-6$

The simplified value of the expression is as follows: $\underline{y\{[\frac{x}{2} - 3] - 4a\} = {}^-6}$

47. Substitute the values for the variables into the expression. $10({}^-5)^3 - 4({}^-5)^2$

Evaluate the power first. $({}^-5)^3 = ({}^-5)({}^-5)({}^-5)] = [({}^-5)({}^-5)]({}^-5)$

$= [25]\,({}^-5)$

$= {}^-125$

$({}^-5)^2 = ({}^-5)({}^-5) = 25$

Substitute. $10[{}^-125) - 4(25)$

Next, multiply as it arises left to right.

Multiply the first two terms. Signs different? Multiply the terms and give the result a negative sign. $[10\ ({}^-125) = {}^-1{,}250]$

Substitute. ${}^-1{,}250 - 4(25)$

Multiply the last two terms. Signs the same? Multiply the terms and give the result a positive sign. $[4(25) = 100]$

Substitute. ${}^-1{,}250 - 100$

Substitute the results into the numerical expression. ${}^-1{,}250 - {}^+100$

Change subtraction to addition and the sign of the term that follows. ${}^-1{,}250 + {}^-100$

Same signs? Add the value of the terms and give the result the same sign. $[1{,}250 + 100 = 1{,}350]$

$10({}^-5)^3 - 4({}^-5)^2 = {}^-1{,}350$

The simplified value of the expression is as follows: $\underline{10b^3 - 4b^2 = {}^-1{,}350}$

48. Substitute the values for the variables into the expression. $-({}^-5 + 6 + \frac{1}{2})^2$

Order of operations tells you first to simply the expressions enclosed within the parentheses.

Add and subtract terms as they arise from left to right. $[{}^-5 + 6 + \frac{1}{2} = ({}^-5 + 6) + \frac{1}{2}]$

$[({}^-5 + 6) + \frac{1}{2} = (1) + \frac{1}{2}]$

$\left[1 + \frac{1}{2} = \frac{2}{2} + \frac{1}{2} = \frac{2 + 1}{2} = \frac{3}{2}\right]$

Substitute. ${}^-(\frac{3}{2})^2$

Next, compute the power. $[(\frac{3}{2})^2 = \frac{3}{2} \cdot \frac{3}{2} = \frac{3 \cdot 3}{2 \cdot 2} = \frac{9}{4}]$

Substitute. ${}^-(\frac{9}{4})$

The original expression now simplifies to: $\underline{{}^-({}^-5 + 6 + \frac{1}{2})^2 = {}^-\frac{9}{4}}$

49. Substitute the values for the variables into the expression. $({}^-8)^2 - 4(3)^2(\frac{1}{2})$

Evaluate the first term. $[({}^-8)^2 = {}^-8 \cdot {}^-8]$

Signs the same? Multiply and give the result a positive sign. $[{}^-8 \cdot {}^-8 = 64]$

Next, in the second expression, evaluate the term with the exponent first. $[(3)^2 = (3)(3) = 9]$

Substitute. $4(9)(\frac{1}{2})$

Multiply the terms from left to right. $[4(9)(\frac{1}{2}) = (4 \cdot 9)(\frac{1}{2}) = 36(\frac{1}{2})]$

$[36(\frac{1}{2}) = \frac{36 \cdot 1}{1 \cdot 2} = 18]$

Substitute the results into the numerical expression. $(64) - (18)$

Now, just subtract. $64 - 18 = 46$

The simplified value of the expression is as follows: $\underline{z^2 - 4a^2y = 46}$

50. Substitute the values for the variables into the expression. $3(6)^2(^-5)[5(3) - 3(^-5)]$

PEMDAS: Evaluate the expression in the brackets first. $[[5(3) - 3(^-5)] = 5 \cdot 3 - 3 \cdot {}^-5]$

$[5 \cdot 3 - 3 \cdot {}^-5 = 15 - {}^-15]$

Change subtraction to addition and the sign of the term that follows. $[15 + {}^+15 = 30]$

Substitute the result into the numerical expression. $3(6)^2(^-5)(30)$

PEMDAS: Evaluate terms with exponents next. $[(6)^2 = 6 \cdot 6 = 36]$

Substitute the result into the numerical expression. $3(36)(^-5)(30)$

Multiply from left to right. $[3(36) = 108]$

Signs different? Multiply the values and give a negative sign. $[(108) \cdot (^-5) = {}^-540]$

$[(^-540) \cdot (30) = {}^-16{,}200]$

$3(6)^2(^-5)[5(3) - 3(^-5)] = {}^-16{,}200$

The simplified value of the expression is as follows: $\underline{3x^2b[5a - 3b] = {}^-16{,}200}$

3

Combining Like Terms

In this chapter, you will practice simplifying algebraic expressions. As you do this, you will recognize and combine terms with variables that are alike and link them to other terms using the arithmetic operations.

You should know that

- the numbers in front of the variable or variables are called coefficients.
- a coefficient is just a factor in an algebraic term, as are the variable or variables in the term.
- like terms can have different coefficients, but the configuration of the variables must be the same for the terms to be alike. For example, $3x$ and $^{-}4x$ are like terms but are different from $7ax$ or $2x^3$.

You can think of an algebraic term as a product of factors consisting of numbers and variables. When the variables are given numerical values, you can multiply the factors of a term together to find its value, as you did in Chapter 2. When you have terms that are alike, you can add or subtract them as if they were signed numbers. You may find that combining like terms may be easier if you do addition by changing all subtraction to addition of the following term with its sign changed. This strategy will continue to be shown in the answer explanations. But as you either know or are beginning to see, sometimes it's easier to just subtract.

You will also use the important commutative and associative properties of addition and multiplication. Another important and useful property is the distributive property. See the **Tips for Combining Like Terms**.

Tips for Combining Like Terms

Distributive Property of Multiplication

The distributive property of multiplication tells you how to multiply the terms inside a parentheses by the term outside the parentheses. Study the following general and specific examples.

$a(b + c) = ab + ac$
$a(b - c) = ab - ac$
$(b + c)a = ba + ca$
$4(6 + 3) = 4 \cdot 6 + 4 \cdot 3 = 24 + 12 = 36$
$(^-5 + 8)3 = {^-5} \cdot 3 + 8 \cdot 3 = {^-15} + 24 = 9$
$7(10 + 3) = 7 \cdot 10 + 7 \cdot 3 = 70 + 21 = 91$
$3(x + 2y) = 3 \cdot x + 3 \cdot 2y = 3x + 6y$
$a(b - 5d) = a \cdot b - a \cdot 5d = ab - 5ad$

Numerical examples of the commutative properties for addition and multiplication were given in the **Tips for Working with Integers**. Now consider the following examples:

Commutative Property of Addition

$a + b = b + a$

This equation reminds us that terms being combined by addition can change their location (commute), but the value of the expression remains the same.

Commutative Property of Multiplication

$x \cdot y = y \cdot x$

This equation reminds us that the order in which we multiply expressions can change without changing the value of the result.

Associative Property of Addition

$(q + r) + s = q + (r + s)$

This equation reminds us that when you are performing a series of additions of terms, you can associate any term with any other and the result will be the same.

Associative Property of Multiplication

$(d \cdot e) \cdot f = d \cdot (e \cdot f)$

This equation reminds us that you can multiply three or more terms in any order without changing the value of the result.

Identity Property of Addition

$n + 0 = n$

Identity Property of Multiplication

$n \cdot 1 = n$

Term Equivalents

$x = 1 \cdot x$

For purposes of combining like terms, a variable by itself is understood to mean one of that term.

$n = {}^{+}1n$

A term without a sign in front of it is considered to be positive.
Adding a negative term is the same as subtracting a positive term.

$a + {}^{-}b = a - {}^{+}b = a - b$

Look at the expressions on either side of the equal signs. Which one looks simplest? Of course, it's the last, $a - b$. Clarity is valued in mathematics. Writing expressions as simply as possible is always appreciated.

While it may not seem relevant yet, as you work through the practice exercises, you will see how each of these properties will come into play as you simplify algebraic expressions by combining like terms.

Simplify the following expressions by combining like terms.

51. $5a + 2a + 7a$

52. $7a + 6b + 3a$

53. $4x + 2y - x + 3y$

54. $27 - 3m + 12 - 5m$

55. $7h + 6 + 2w - 3 + h$

56. $4(x + 2y) + 2(x + y)$

57. $3(2a + 3b) + 7(a - b)$

58. $11(4m + 5) + 3(^-3m + 8)$

59. $64 + 5(n - 8) + 12n - 24$

60. $4(x + y - 4) + 6(2 - 3y)$

61. $^-3(2z - 3w) + 2(2w - 4 - 3z)$

62. $14 + 9(2w + 7) - 2(6 - w)$

63. $(3 - 2s - 5t) - 2(3t - s - 1) + 5$

64. $^-m(2 - 4n) + 2n(^-1 + 3m)$

65. $w(^-w - z + 2) + z(4z - 2w)$

66. $a(a + 4) + 3a^2 - 2a + 10$

67. $wz - 2z(w - 3z - 1) - 5(wz + z^2)$

68. $3r^2 + r(2 - r) + 6(r + 4)$

69. $2x - x(5 + y) + 3xy + 4(2x - y)$

70. $ab(^-1 - 2c) - 3a(b + bc + 1) + b(^-c + 2a)$

71. $5(3x - y) + x(5 + 2y) - 4(3 + x)$

72. $6(m - 3n) + 3m(n + 5) - 2n(3 - m)$

73. $9(2x - t) + 23xt + x(^-4 + 5t)$

74. $4\{2a(a + 3) + 6(4 - a)\} + 5a^2$

75. $8(2a - b - 3c) + 3(2a - b) - 4(6 - b)$

Answers

Numerical expressions in parentheses like this [ ] are operations performed on only part of the original expression. The operations performed within these symbols are intended to show how to evaluate the various terms that make up the entire expression.

Expressions with parentheses that look like this ( ) contain either numerical substitutions or expressions that are part of a numerical expression. Once a single number appears within these parentheses, the parentheses are no longer needed and need not be used the next time the entire expression is written.

When two pair of parentheses appear side by side like this ( )( ), it means that the expressions within are to be multiplied.

Sometimes parentheses appear within other parentheses in numerical or algebraic expressions. Regardless of what symbol is used, ( ), { }, or [ ], perform operations in the innermost parentheses first and work outward.

Underlined expressions show the simplified result.

51. Use the associative property of addition. $(5a + 2a) + 7a$
Add like terms. $[5a + 2a = 7a]$
Substitute the results into the original expression. $(7a) + 7a$
Add like terms. $7a + 7a = 14a$
The simplified result of the algebraic expression is: $\underline{14a}$

52. Use the commutative property of addition to move like terms together. $7a + 3a + 6b$
Use the associative property for addition. $(7a + 3a) + 6b$
Add like terms. $[(7a + 3a) = 10a]$
Substitute. $(10a) + 6b$
The simplified result of the algebraic expression is: $\underline{10a + 6b}$

53. Change subtraction to addition and change the sign of the term that follows. $4x + 2y + (^-x) + 3y$
Use the commutative property of addition to move like terms together. $4x + (^-x) + 2y + 3y$
Use the associative property for addition. $(4x + {}^-x) + (2y + 3y)$
Add like terms. $[4x + {}^-x = {}^+3x = 3x]$
$[2y + 3y = 5y]$
Substitute the results into the expression. $(4x + {}^-x) + (2y + 3y) = (3x) + (5y)$
The simplified algebraic expression is: $\underline{3x + 5y}$

54. Change subtraction to addition and change the sign of the term that follows. $27 + {}^{-}3m + 12 + {}^{-}5m$

Use the commutative property for addition to put like terms together. $27 + 12 + {}^{-}3m + {}^{-}5m$

Use the associative property for addition. $(27 + 12) + ({}^{-}3m + {}^{-}5m)$

Add like terms. $[27 + 12 = 39]$
$[{}^{-}3m + {}^{-}5m = {}^{-}8m]$

Substitute the results into the expression. $(27 + 12) + ({}^{-}3m + {}^{-}5m) = (39) + ({}^{-}8m)$

Rewrite addition of a negative term as subtraction of a positive term by changing addition to subtraction and changing the sign of the following term. $39 - {}^{+}8m = 39 - 8m$

The simplified algebraic expression is: $\underline{39 - 8m}$

55. Change subtraction to addition and change the sign of the term that follows. $7h + 6 + 2w + ({}^{-}3) + h$

Use the commutative property for addition to put like terms together. $7h + h + 2w + 6 + {}^{-}3$

Use the associative property for addition. $(7h + h) + 2w + (6 + {}^{-}3)$

Add like terms. $[(7h + h) = 8h]$
$[(6 + {}^{-}3) = 3]$

Substitute the result into the expression. $(8h) + 2w + (3)$

The simplified algebraic expression is: $\underline{8h + 2w + 3}$

56. Use the distributive property of multiplication on the first expression. $[4(x + 2y) = 4 \cdot x + 4 \cdot 2y]$
$[4x + 8y]$

Use the distributive property of multiplication on the second expression. $[2(x + y) = 2 \cdot x + 2 \cdot y]$
$[2x + 2y]$

Substitute the results into the expression. $(4x + 8y) + (2x + 2y)$

Use the commutative property of addition to put like terms together. $4x + 2x + 8y + 2y$

Use the associative property for addition. $(4x + 2x) + (8y + 2y)$

Add like terms. $[4x + 2x = 6x]$
$[8y + 2y = 10y]$

Substitute the results into the expression. $(6x) + (10y)$

The simplified algebraic expression is: $\underline{6x + 10y}$

57. Use the distributive property of multiplication on the first term. $[3(2a + 3b) = 3 \cdot 2a + 3 \cdot 3b]$ $[6a + 9b]$

Use the distributive property of multiplication on the second term. $[7(a - b) = 7 \cdot a - 7 \cdot b]$ $[7a - 7b]$

Substitute the results into the expression. $(6a + 9b) + (7a - 7b)$ $6a + 9b + 7a - 7b$

Change subtraction to addition and change the sign of the term that follows. $6a + 9b + 7a + (^-7b)$
Use the commutative property for addition to put like terms together. $6a + 7a + 9b + (^-7b)$
Use the associative property for addition. $(6a + 7a) + (9b + {^-7b})$
Add like terms. $[6a + 7a = 13a]$
Signs different? Subtract the value of the terms. $[9b + {^-7b} = 2b]$
Substitute the result into the expression. $(13a) + (2b)$
The simplified algebraic expression is: $\underline{13a + 2b}$

58. Use the distributive property of multiplication on the first term. $[11(4m + 5) = 11 \cdot 4m + 11 \cdot 5]$ $[44m + 55]$

Use the distributive property of multiplication on the second term. $[3(^-3m + 8) = 3 \cdot (^-3m) + 3 \cdot 8]$ $[^-9m + 24]$

Substitute the result into the expression. $(44m + 55) + (^-9m + 24)$ $44m + 55 + {^-9m} + 24$

Use the commutative property for addition to put like terms together. $44m + {^-9m} + 55 + 24$
Use the associative property for addition. $(44m + {^-9m}) + (55 + 24)$
Add like terms. $[44m + {^-9m} = 35m]$ $[55 + 24 = 79]$

Substitute the result into the expression. $(35m) + (79)$
The simplified algebraic expression is: $\underline{35m + 79}$

59. Use the distributive property of multiplication on the second term. $[5(n - 8) = 5 \cdot n - 5 \cdot 8]$ $[5n - 40]$

Substitute the result into the expression. $64 + (5n - 40) + 12n - 24$
Parentheses are no longer needed. $64 + 5n - 40 + 12n - 24$
Change subtraction to addition and change the sign of the term that follows. $64 + 5n + {^-40} + 12n + {^-24}$

Use the commutative property for addition to put like terms together.	$5n + 12n + 64 + {}^-40 + {}^-24$
Use the associative property for addition.	$(5n + 12n) + (64 + {}^-40 + {}^-24)$
Add like terms.	$[5n + 12n = 17n]$
Add like terms.	$[64 + {}^-40 + {}^-24 = 64 + ({}^-40 + {}^-24)]$ $[64 + {}^-64 = 0]$
Substitute the results into the expression.	$(17n) + (0)$
The simplified algebraic expression is:	$\underline{17n}$

60.

Use the distributive property of multiplication on the first term.	$[4(x + y - 4) = 4 \cdot x + 4 \cdot y - 4 \cdot 4]$ $[4 \cdot x + 4 \cdot y - 4 \cdot 4 = 4x + 4y - 16]$
Use the distributive property of multiplication on the second term.	$[6(2 - 3y) = 6 \cdot 2 - 6 \cdot 3y = 12 - 18y]$
Substitute the results into the expression.	$(4x + 4y - 16) + (12 - 18y)$
Parentheses are no longer needed.	$4x + 4y - 16 + 12 - 18y$
Use the commutative property for addition to put like terms together.	$4x + 4y - 18 - 16 + 12$
Change subtraction to addition and change the sign of the terms that follow.	$4x + 4y + {}^-18y + {}^-16 + 12$
Use the associative property for addition.	$4x + (4y + {}^-18y) + ({}^-16 + 12)$
Add like terms.	$[4y + {}^-18y = {}^-14y]$ $[{}^-16 + 12 = {}^-4]$
Substitute the results into the expression.	$4x + ({}^-14y) + ({}^-4)$
Rewrite addition of a negative term as subtraction of a positive term by changing addition to subtraction and changing the sign of the following term.	
The simplified algebraic expression is:	$\underline{4x - 14y - 4}$

61. Use the distributive property of multiplication on the first term. $[{}^{-}3(2z - 3w) = {}^{-}3(2z) - ({}^{-}3)(3w)]$ $[{}^{-}3(2z) - ({}^{-}3)(3w) = {}^{-}6z - ({}^{-}9w) = {}^{-}6z + 9w]$

Use the distributive property of multiplication on the second term. $[2(2w - 4 - 3z) = 2(2w) - 2(4) - 2(3z)]$ $[2(2w) - 2(4) - 2(3z) = 4w - 8 - 6z]$

Substitute the results into the original expression. $({}^{-}6z + 9w) + (4w - 8 - 6z)$

Parentheses are no longer needed. ${}^{-}6z + 9w + 4w - 8 - 6z$

Use the commutative property of addition to put like terms together. ${}^{-}6z - 6z + 9w + 4w - 8$

Use the associative property of addition. $({}^{-}6z - 6z) + (9w + 4w) - 8$

Add like terms. $[{}^{-}6z - 6z = {}^{-}12z]$ $[9w + 4w = 13w]$

Substitute the results into the expression. $({}^{-}12z) + (13w) - 8$

The simplified algebraic expression is: $\underline{{}^{-}12z + 13w - 8}$

62. Change subtraction to addition and change the sign of the terms that follow. $14 + 9(2w + 7) + {}^{-}2(6 + {}^{-}w)$

Use the distributive property of multiplication on the second term. $[9(2w + 7) = 9 \cdot 2w + 9 \cdot 7]$ $[9 \cdot 2w + 9 \cdot 7 = 18w + 63]$

Use the distributive property of multiplication on the third term. $[{}^{-}2(6 + {}^{-}w) = {}^{-}2 \cdot 6 + {}^{-}2 \cdot {}^{-}w]$

Notice the result of multiplication for opposite and like-signed terms. $[{}^{-}2 \cdot 6 + {}^{-}2 \cdot {}^{-}w = {}^{-}12 + {}^{+}2w]$

Substitute the results into the original expression. $14 + (18w + 63) + ({}^{-}12 + {}^{+}2w)$

Parentheses are no longer needed. $14 + 18w + 63 + {}^{-}12 + {}^{+}2w$

Use the commutative property of addition to put like terms together. $18w + {}^{+}2w + 14 + 63 + {}^{-}12$

Use the associative property for addition. $(18w + {}^{+}2w) + (14 + 63 + {}^{-}12)$

Add like terms. $[18w + {}^{+}2w = 20w]$

Add from left to right. $[14 + 63 + {}^{-}12 = 77 + {}^{-}12 = 65]$

Substitute the results into the expression. $(20w) + (65)$

The simplified algebraic expression is: $\underline{20w + 65}$

63. Use the distributive property of multiplication on the second term. $^-2(3t - s - 1) = {^-2}(3t) - 2(^-s) - 2(^-1)$
$= {^-6t} + 2s + 2$

Substitute the result into the original expression. $(3 - 2s - 5t) + (^-6t + 2s + 2) + 5$

Parentheses are no longer needed. $3 - 2s - 5t - 6t + 2s + 2 + 5$

Use the commutative property of addition to put like terms together. $^-2s + 2s - 5t - 6t + 2 + 5 + 3$

Use the associative property of addition. $(^-2s + 2s) + (^-5t - 6t) + (2 + 5 + 3)$

Add like terms. $0s - 11t + 10$

The simplified algebraic expression is: $\underline{^-11t + 10}$

64. Use the distributive property of multiplication on the first term. $^-m(2 - 4n) = {^-m}(2) - m(^-4n)$
$= {^-2m} + 4mn$

Use the distributive property of multiplication on the second term. $2n(^-1 + 3m) = 2n(^-1) + 2n(3m)$
$= {^-2n} + 6mn$

Substitute the results into the original expression. $(^-2m + 4mn) + (^-2n + 6mn)$

Parentheses are no longer needed. $^-2m + 4mn - 2n + 6mn$

Use the commutative property of addition to put like terms together. $^-2m - 2n + 6mn + 4mn$

Use the associative property of addition. $^-2m - 2n + (6mn + 4mn)$

Add like terms. $^-2m - 2n + (10mn)$

The simplified algebraic expression is: $\underline{^-2m - 2n + 10mn}$

65. Use the distributive property of multiplication on the first term. $w(^-w - z + 2) = w(^-w) + w(^-z) + w(2)$
$= {^-w^2} - wz + 2w$

Use the distributive property of multiplication on the second term. $z(4z - 2w) = z(4z) + z(^-2w)$
$= 4z^2 - 2wz$

Substitute the results into the original expression. $(^-w^2 - wz + 2w) + (4z^2 - 2wz)$

Parentheses are no longer needed. $^-w^2 - wz + 2w + 4z^2 - 2wz$

Use the commutative property of addition to put like terms together. $^-w^2 + 4z^2 - wz - 2wz + 2w$

Use the associative property of addition. $^-w^2 + 4z^2 + (^-wz - 2wz) + 2w$

Add like terms. $^-w^2 + 4z^2 + (^-3wz) + 2w$

The simplified algebraic expression is: $\underline{^-w^2 + 4z^2 - 3wz + 2w}$

66. Change subtraction to addition and change the sign of the terms that follow. $a(a + 4) + 3a^2 + {}^-2a + 10$

Use the distributive property of multiplication on the first term. $[a(a + 4) = a \cdot a + a \cdot 4]$

Use the commutative property for multiplication for the second term. $[a \cdot a + a \cdot 4 = a^2 + 4a]$

Substitute the results into the expression. $(a^2 + 4a) + 3a^2 + {}^-2a + 10$

Parentheses are no longer needed. $a^2 + 4a + 3a^2 + {}^-2a + 10$

Use the commutative property of addition to put like terms together. $a^2 + 3a^2 + 4a + {}^-2a + 10$

Use the associative property for addition. $(a^2 + 3a^2) + (4a + {}^-2a) + 10$

Add the first term using the rules for terms with the same signs. $[a^2 + 3a^2 = 4a^2]$

Add the second term using the rules for terms with different signs. $[4a + {}^-2a = 2a]$

Substitute the results into the expression. $(4a^2) + (2a) + 10$

The simplified algebraic expression is: $\underline{4a^2 + 2a + 10}$

67. Use the distributive property of multiplication on the second term. ${}^-2z(w - 3z - 1) = {}^-2z(w) - 2z({}^-3z) - 2z({}^-1)$
$= {}^-2wz + 6z^2 + 2z$

Use the distributive property of multiplication on the third term. ${}^-5(wz + z^2) = {}^-5wz - 5z^2$

Substitute the results into the original expression. $wz + ({}^-2wz + 6z^2 + 2z) + ({}^-5wz - 5z^2)$

Parentheses are no longer needed. $wz - 2wz + 6z^2 + 2z - 5wz - 5z^2$

Use the commutative property of addition to put like terms together. $wz - 2wz - 5wz + 6z^2 - 5z^2 + 2z$

Use the associative property of addition. $(wz - 2wz - 5wz) + (6z^2 - 5z^2) + 2z$

Add like terms. $({}^-6wz) + (z^2) + 2z$

The simplified algebraic expression is: $\underline{{}^-6wz + z^2 + 2z}$

68. Change subtraction to addition and change the sign of the terms that follow. $3r^2 + r(2 + {}^-r) + 6(r + 4)$

Use the distributive property of multiplication on the second term. $[r(2 + {}^-r) = r \cdot 2 + r \cdot {}^-r]$
$[r \cdot 2 + r \cdot {}^-r = 2r + {}^-r^2]$

Use the distributive property of multiplication on the third term. $[6(r + 4) = 6 \cdot r + 6 \cdot 4]$
$[6 \cdot r + 6 \cdot 4 = 6r + 24]$

Substitute the results into the expression. $3r^2 + (2r + {}^-r^2) + (6r + 24)$

Remove the parentheses. $3r^2 + 2r + {}^-r^2 + 6r + 24$

Use the commutative property of addition to put like terms together. $3r^2 + {}^{-}r^2 + 2r + 6r + 24$

Use the associative property for addition. $(3r^2 + {}^{-}r^2) + (2r + 6r) + 24$

Add the first term using the rules for terms with different signs. $[3r^2 + {}^{-}r^2 = 2r^2]$

Add the second term using the rules for terms with the same signs. $[2r + 6r = 8r]$

Substitute the results into the expression. $(2r^2) + (8r) + 24$

The simplified algebraic expression is: $\underline{2r^2 + 8r + 24}$

69. Change subtraction to addition and the sign of the terms that follow. $2x + {}^{-}x(5 + y) + 3xy + 4(2x + {}^{-}y)$

Use the distributive property of multiplication on the second term. $[{}^{-}x(5 + y) = {}^{-}x \cdot 5 + {}^{-}x \cdot y]$

Use the rules for multiplying signed terms. $[{}^{-}x \cdot 5 + {}^{-}x \cdot y = {}^{-}5x + {}^{-}xy]$

Use the distributive property of multiplication on the fourth term. $[4(2x + {}^{-}y) = 4 \cdot 2x + 4 \cdot {}^{-}y]$

Use the rules for multiplying signed terms. $[4 \cdot 2x + 4 \cdot {}^{-}y = 8x + {}^{-}4y]$

Substitute the results into the expression. $2x + ({}^{-}5x + {}^{-}xy) + 3xy + (8x + {}^{-}4y)$

Remove the parentheses. $2x + {}^{-}5x + {}^{-}xy + 3xy + 8x + {}^{-}4y$

Use the associative and commutative properties for addition. $(2x + {}^{-}5x + 8x) + ({}^{-}xy + 3xy) + ({}^{-}4y)$

Add the first set of terms using the rules for terms with different signs. $[2x + {}^{-}5x + 8x = 5x]$

Add the second set of terms using the rules for terms with different signs. $[{}^{-}xy + 3xy = 2xy]$

Substitute the results into the expression. $({}^{-}5x) + (2xy) + ({}^{-}4y)$

The simplified algebraic expressions is: $\underline{{}^{-}5x + 2xy - 4y}$

70. Use the distributive property of multiplication on the first term. $ab({}^{-}1 - 2c) = ab({}^{-}1) + ab({}^{-}2c)$
$= {}^{-}ab - 2abc$

Use the distributive property of multiplication on the second term. ${}^{-}3a(b + bc + 1) = {}^{-}3a(b) - 3a(bc) - 3a(1)$
$= {}^{-}3ab - 3abc - 3a$

Use the distributive property of multiplication on the third term. $b({}^{-}c + 2a) = b({}^{-}c) + b(2a)$
$= {}^{-}bc + 2ab$

Substitute the results into the original expression. $({}^{-}ab - 2abc) + ({}^{-}3ab - 3abc - 3a) + ({}^{-}bc + 2ab)$

Parentheses are no longer needed. ${}^{-}ab - 2abc - 3ab - 3abc - 3a - bc + 2ab$

Use the commutative property of addition to put like terms together. ${}^{-}3a - bc - ab - 3ab + 2ab - 2abc - 3abc$

Use the associative property of addition. ${}^{-}3a - bc + ({}^{-}ab - 3ab + 2ab) + ({}^{-}2abc - 3abc)$

Add like terms.	$^{-}3a - bc + (^{-}2ab) + (^{-}5abc)$
The simplified algebraic expression is:	$\underline{^{-}3a - bc - 2ab - 5abc}$

71.

Change subtraction to addition and the sign of the terms that follow.	$5(3x + {}^{-}y) + x(5 + 2y) + {}^{-}4(3 + x)$
Use the distributive property of multiplication on the first term.	$[5(3x + {}^{-}y) = 5 \cdot 3x + 5 \cdot {}^{-}y]$
Use the rules for multiplying signed terms.	$[5 \cdot 3x + 5 \cdot {}^{-}y = 15x + {}^{-}5y]$
Use the distributive property of multiplication on the second term.	$[x(5 + 2y) = x \cdot 5 + x \cdot 2y]$
Use the rules for multiplying signed terms.	$[x \cdot 5 + x \cdot 2y = 5x + 2xy]$
Use the distributive property of multiplication on the third term.	$[{}^{-}4(3 + x) = {}^{-}4 \cdot 3 + {}^{-}4 \cdot x]$
Use the rules for multiplying signed terms.	$[{}^{-}4 \cdot 3 + {}^{-}4 \cdot x = {}^{-}12 + {}^{-}4x]$
Substitute the results into the original expression.	$(15x + {}^{-}5y) + (5x + 2xy) + ({}^{-}12 + {}^{-}4x)$
Remove the parentheses.	$15x + {}^{-}5y + 5x + 2xy + {}^{-}12 + {}^{-}4x$
Use the commutative property of addition to move like terms together. Use the associative property for addition.	$(15x + 5x + {}^{-}4x) + {}^{-}5y + 2xy + {}^{-}12$
Combine like terms using addition rules for signed numbers.	$(16x) + {}^{-}5y + 2xy + {}^{-}12$
Adding a negative term is the same as subtracting a positive term.	$(16x) - ({}^{+}5y) + 2xy - ({}^{+}12)$
The simplified algebraic expression is:	$\underline{16x - 5y + 2xy - 12}$

72. Change subtraction to addition and the sign of the terms that follow. $6(m + {}^-3n) + 3m(n + 5) + {}^-2n(3 + {}^-m)$

Use the distributive property of multiplication on the first term. $[6(m + {}^-3n) = 6 \cdot m + 6 \cdot {}^-3n]$

Use the rules for multiplying signed terms. $[6 \cdot m + 6 \cdot {}^-3n = 6m + {}^-18n]$

Use the distributive property of multiplication on the second term. $[3m(n + 5) = 3m \cdot n + 3m \cdot 5]$

Use the rules for multiplying signed terms. $[3m \cdot n + 3m \cdot 5 = 3mn + 15m]$

Use the distributive property of multiplication on the third term. $[{}^-2n(3 + {}^-m) = {}^-2n \cdot 3 + {}^-2n \cdot {}^-m]$

Use the rules for multiplying signed terms. $[{}^-2n \cdot 3 + {}^-2n \cdot {}^-m = {}^-6n + {}^+2mn]$

Substitute the results into the original expression. $(6m + {}^-18n) + (3mn + 15m) + ({}^-6n + {}^+2mn)$

Remove the parentheses. $6m + {}^-18n + 3mn + 15m + {}^-6n + {}^+2mn$

Use the commutative property of addition to move like terms together. Use the associative property for addition. $(6m + 15m) + (3mn + {}^+2mn) + ({}^-6n + {}^-18n)$

Combine like terms using addition rules for signed numbers. $(21m) + (5mn) + ({}^-24n)$

Adding a negative term is the same as subtracting a positive term. $\underline{21m + 5mn - 24n}$

73. Change subtraction to addition and the sign of the terms that follow. $9(2x + {}^-t) + 23xt + x({}^-4 + 5t)$

Use the distributive property of multiplication on the first term. $[9(2x + {}^-t) = 9 \cdot 2x + 9 \cdot {}^-t]$

Use the rules for multiplying signed terms. $[9 \cdot 2x + 9 \cdot {}^-t = 18x + {}^-9t]$

Use the distributive property of multiplication on the third term. $[x({}^-4 + 5t) = x \cdot {}^-4 + x \cdot 5t]$

Use the rules for multiplying signed terms. $[x \cdot {}^-4 + x \cdot 5t = {}^-4x + 5xt]$

Substitute the results into the expression. $(18x + {}^-9t) + 23xt + ({}^-4x + 5xt)$

Remove the parentheses. $18x + {}^-9t + 23xt + {}^-4x + 5xt$

Use the commutative property of addition to move like terms together. $18x + {}^{-}4x + {}^{-}9t + 23xt + 5xt$

Use the associative property for addition. $(18x + {}^{-}4x) + {}^{-}9t + (23xt + 5xt)$

Combine like terms using addition rules for signed numbers. $(14x) + {}^{-}9t + (28xt)$

Adding a negative term is the same as subtracting a positive term. $14x - {}^{+}9t + 28xt$

$\underline{14x - 9t + 28xt}$

74. Change subtraction to addition and the sign of the terms that follow. $4\{2a(a + 3) + 6(4 + {}^{-}a)\} + 5a^2$

Simplify the term inside the curly brackets first. $[2a(a + 3) + 6(4 + {}^{-}a)]$

Use the distributive property of multiplication on the first term. $[2a(a + 3) = 2a \cdot a + 2a \cdot 3]$

Use the rules for multiplying signed terms. $[2a \cdot a + 2a \cdot 3 = 2a^2 + 6a]$

Use the distributive property of multiplication on the second term. $[6(4 + {}^{-}a) = 6 \cdot 4 + 6 \cdot {}^{-}a]$

Use the rules for multiplying signed terms. $[6 \cdot 4 + 6 \cdot {}^{-}a = 24 + {}^{-}6a]$

Substitute the results into the expression. $[(2a^2 + 6a) + (24 + {}^{-}6a)]$

Remove the parentheses. $[2a^2 + 6a + 24 + {}^{-}6a]$

Use the commutative property of addition. $[2a^2 + 6a + {}^{-}6a + 24]$

Use the associative property for addition. $[2a^2 + (6a + {}^{-}6a) + 24]$

Combine like terms using addition rules for signed numbers. $[2a^2 + (0) + 24]$

Use the identity property of addition. $[2a^2 + 24]$

Substitute the results into the expression. $4\{2a^2 + 24\} + 5a^2$

Use the distributive property of multiplication on the first term. $[4 \cdot 2a^2 + 4 \cdot 24]$

$[8a^2 + 96]$

Substitute into the expression. $(8a^2 + 96) + 5a^2$

Remove the parentheses. $8a^2 + 96 + 5a^2$

Use the commutative property of addition. $8a^2 + 5a^2 + 96$

Use the associative property for addition. $(8a^2 + 5a^2) + 96$

Add like terms. $\underline{13a^2 + 96}$

75.

Change subtraction to addition and the sign of the terms that follow.	$8(2a + {}^{-}b + {}^{-}3c) + 3(2a + {}^{-}b) + {}^{-}4(6 + {}^{-}b)$
Use the distributive property of multiplication on the first term.	$[8(2a + {}^{-}b + {}^{-}3c) = 8 \cdot 2a + 8 \cdot {}^{-}b + 8 \cdot {}^{-}3c]$
Use the rules for multiplying signed terms.	$[8 \cdot 2a + 8 \cdot {}^{-}b + 8 \cdot {}^{-}3c = 16a + {}^{-}8b + {}^{-}24c]$
Use the distributive property of multiplication on the second term.	$[3(2a + {}^{-}b) = 3 \cdot 2a + 3 \cdot {}^{-}b]$
Use the rules for multiplying signed terms.	$[3 \cdot 2a + 3 \cdot {}^{-}b = 6a + {}^{-}3b]$
Use the distributive property of multiplication on the third term.	$[{}^{-}4(6 + {}^{-}b) = {}^{-}4 \cdot 6 + {}^{-}4 \cdot {}^{-}b]$
Use the rules for multiplying signed terms.	$[{}^{-}4 \cdot 6 + {}^{-}4 \cdot {}^{-}b = {}^{-}24 + {}^{+}4b]$
Substitute the results into the expression.	$(16a + {}^{-}8b + {}^{-}24c) + (6a + {}^{-}3b) + ({}^{-}24 + {}^{+}4b)$
Remove the parentheses.	$16a + {}^{-}8b + {}^{-}24c + 6a + {}^{-}3b + {}^{-}24 + {}^{+}4b$
Use the commutative property of addition to move like terms together.	$16a + 6a + {}^{-}8b + {}^{-}3b + {}^{+}4b + {}^{-}24c + {}^{-}24$
Use the associative property for addition.	$(16a + 6a) + ({}^{-}8b + {}^{-}3b + {}^{+}4b) + {}^{-}24c + {}^{-}24$
Combine like terms using addition rules for signed numbers.	$(22a) + ({}^{-}7b) + {}^{-}24c + {}^{-}24$
Adding a negative term is the same as subtracting a positive term.	$\underline{22a - 7b - 24c - 24}$

4

Solving Basic Equations

Solving equations is not very different from working with numerical or algebraic expressions. An equation is a mathematical statement where two expressions are set equal to each other. Using logic and mathematical operations, you can manipulate the terms of the equation to find its solution. Simply put, that is what you have done when you have the variable on one side of the equal sign and a number on the other. The answer explanations will show and identify all the steps used to solve basic equations. There will be different solutions to similar problems to illustrate a variety of methods for solving equations. But they all rely on the same rules. Look over the **Tips for Solving Basic Equations** before you begin this chapter's questions.

Tips for Solving Basic Equations

- If a number is being added to or subtracted from a term on one side of an equation, you can eliminate that number by performing the inverse operation.
- The inverse of addition is subtraction. The inverse of subtraction is addition. If you add or subtract a quantity from one side of the equation, you must do the same to the other side to maintain the equality.

- The inverse of multiplication is division. If a variable is being multiplied by a coefficient, you can eliminate the coefficient by dividing both sides of the equation by that coefficient, leaving you with just one of the variables. If a is a number $\neq 0$, then $ax \div a = x$.
- The inverse of division is multiplication. If a number is dividing a variable, you can multiply the term by the number, leaving you with one of the variables. If b is a number $\neq 0$, then $b(\frac{y}{b}) = y$.
- When you have the variable isolated on one side of the equation, the value on the other side is the solution.

For numbers 76–95, solve the equation.

76. $a + 21 = 32$

77. $x - 25 = 32$

78. $y + 17 = {}^-12$

79. $b - 23 = {}^-18$

80. $c - {}^-3 = {}^-3$

81. $s - {}^-4 = {}^-1$

82. $a + \frac{5}{6} = {}^-\frac{1}{6}$

83. $b - \frac{5}{2} = \frac{-2}{3}$

84. $c - 0.30 = {}^-2.29$

85. $d - 1.016 = {}^-1.016$

86. $2a = 24$

87. $4x = {}^-20$

88. ${}^-3y = 18$

89. ${}^-27b = {}^-9$

90. $45r = {}^-30$

91. $0.2c = 5.8$

92. $\frac{x}{7} = 16$

93. $\frac{y}{-4} = {}^-12$

94. $\frac{2}{3}a = 54$

95. $\frac{8}{5}b = -\frac{56}{15}$

For numbers 96–100, solve the word problem.

96. Jack paid \$28,000 for his new car. This was $\frac{7}{8}$ the suggested selling price of the car. What was the suggested selling price of the car?

97. After putting 324 teddy bears into packing crates, there were 54 crates filled with bears. If each crate contained the same number of bears, how many bears were in each packing crate?

98. Only 3% of turtle hatchlings will live to become breeding adults. How many turtles must have been born if the current number of breeding adults is 1,200?

99. This year, a farmer planted 300 acres of corn. This was 1.5 times as many acres as he planted last year. How many acres did he plant last year?

100. A business executive received a \$6,000 bonus check from her company at the end of the year. This was 5% of her annual salary. How much was her annual salary before receiving the bonus?

Answers

Numerical expressions in parentheses like this [ ] are operations performed on only part of the original expression. The operations performed within these symbols are intended to show how to evaluate the various terms that make up the entire expression.

Expressions with parentheses that look like this ( ) contain either numerical substitutions or expressions that are part of a numerical expression. Once a single number appears within these parentheses, the parentheses are no longer needed and need not be used the next time the entire expression is written.

When two pair of parentheses appear side by side like this ( )( ), it means that the expressions within are to be multiplied.

Sometimes parentheses appear within other parentheses in numerical or algebraic expressions. Regardless of what symbol is used, ( ), { }, or [ ], perform operations in the innermost parentheses first and work outward.

Underlined equations show the simplified result.

76. Subtract 21 from both sides of the equation. $a + 21 - 21 = 32 - 21$
Associate like terms. $a + (21 - 21) = (32 - 21)$
Perform the numerical operation in the parentheses. $a + (0) = (11)$
Zero is the identity element for addition. $\underline{a = 11}$

77. Add 25 to each side of the equation. $25 + x - 25 = 32 + 25$
Use the commutative property for addition. $x + 25 - 25 = 32 + 25$
Associate like terms. $x + (25 - 25) = (32 + 25)$
Perform the numerical operation in the parentheses. $x + (0) = (57)$
Zero is the identity element for addition. $\underline{x = 57}$

78. Subtract 17 from both sides of the equation. $y + 17 - 17 = {}^{-}12 - 17$
Associate like terms. $y + (17 - 17) = ({}^{-}12 - 17)$
Change subtraction to addition and change the sign of the term that follows. $y + (17 + {}^{-}17) = ({}^{-}12 + {}^{-}17)$
Apply the rules for operating with signed numbers. $y + (0) = {}^{-}29$
Zero is the identity element for addition. $\underline{y = {}^{-}29}$

79. Change subtraction to addition and change the sign of the term that follows. $b + {}^{-}23 = {}^{-}18$
Add ${}^{+}23$ to each side of the equation. $b + {}^{-}23 + 23 = {}^{-}18 + 23$
Associate like terms. $b + ({}^{-}23 + 23) = ({}^{-}18 + 23)$
Apply the rules for operating with signed numbers. $b + (0) = (5)$
Zero is the identity element for addition. $\underline{b = 5}$

80. Use the rules for operating with signed numbers to simplify the left side of the equation. $c + {}^{+}3 = {}^{-}3$

$c + 3 = {}^{-}3$

Subtract 3 from each side of the equation. $c + 3 - 3 = {}^{-}3 - 3$

Associate like terms. $c + (3 - 3) = ({}^{-}3 - 3)$

Apply the rules for operating with signed numbers. $c + (0) = ({}^{-}6)$

Zero is the identity element for addition. $\underline{c = {}^{-}6}$

81. Change subtraction to addition and change the sign of the term that follows. $s + {}^{+}4 = {}^{-}1$

Add ⁻4 to each side of the equation. $s + {}^{+}4 + {}^{-}4 = {}^{-}1 + {}^{-}4$

Associate like terms. $s + ({}^{+}4 + {}^{-}4) = {}^{-}1 + {}^{-}4$

Apply the rules for operating with signed numbers. $s + (0) = ({}^{-}5)$

Subtracting zero is the same as adding zero. $\underline{s = {}^{-}5}$

82. Subtract $\frac{5}{6}$ from each side of the equation. $a + \frac{5}{6} - \frac{5}{6} = {}^{-}\frac{1}{6} - \frac{5}{6}$

Associate like terms. $a + (\frac{5}{6} - \frac{5}{6}) = ({}^{-}\frac{1}{6} - \frac{5}{6})$

Apply the rules for operating with singed numbers. $a + (0) = ({}^{-}1)$

Zero is the identity element for addition. $\underline{a = {}^{-}1}$

83. Add $\frac{5}{2}$ to each side of the equation. $b - \frac{5}{2} + \frac{5}{2} = {}^{-}\frac{2}{3} + \frac{5}{2}$

Change subtraction to addition and change the sign of the term that follows. $b + {}^{-}\frac{5}{2} + \frac{5}{2} = {}^{-}\frac{2}{3} + \frac{5}{2}$

Associate like terms. $b + ({}^{-}\frac{5}{2} + \frac{5}{2}) = ({}^{-}\frac{2}{3} + \frac{5}{2})$

Apply the rules for operating with signed numbers. $b + (0) = \frac{11}{6}$

Change the improper fraction to a mixed number. $\underline{b = \frac{11}{6} = 1\frac{5}{6}}$

84. Change subtraction to addition and change the sign of the term that follows. $c + {}^{-}0.30 = {}^{-}2.29$

Add ${}^{+}0.30$ to each side of the equation. $c + {}^{-}0.30 + 0.30 = {}^{-}2.29 + 0.30$

Associate like terms. $c + ({}^{-}0.30 + 0.30) = ({}^{-}2.29 + 0.30)$

Apply the rules for operating with signed numbers. $c + (0) = ({}^{-}1.99)$

Zero is the identity element for addition. $\underline{c = {}^{-}1.99}$

85. Subtract 1.016 from each side of the equation. $d + 1.016 - 1.016 = {}^{-}1.016 - 1.016$

Associate like terms. $d + (1.016 - 1.016) = ({}^{-}1.016 - 1.016)$

Apply the rules for operating with signed numbers. $d + (0) = ({}^{-}2.032)$

Zero is the identity element for addition. $\underline{d = {}^{-}2.032}$

86. Divide both sides of the equation by 2. $\frac{2a}{2} = \frac{24}{2}$

Apply the rules for operating with signed numbers. $\underline{a = 12}$

87. Divide both sides of the equation by 4. $\frac{4x}{4} = \frac{-20}{4}$

Apply the rules for operating with signed numbers. $\underline{x = {}^{-}5}$

88. Divide both sides of the equation by ${}^{-}3$. $\frac{-3y}{-3} = \frac{18}{-3}$

$y = \frac{18}{-3}$

Apply the rules for operating with signed numbers. $y = {}^{-}6$

89. Divide both sides of the equation by ${}^{-}27$. $\frac{-27b}{-27} = \frac{-9}{-27}$

Reduce fractions to their simplest form (common factor of 9). Signs the same? $b = \frac{-9}{-27}$

Give the result a positive sign. $\underline{b = {}^{+}\frac{1}{3}}$

90. Divide both sides of the equation by 45. $\frac{45r}{45} = \frac{-30}{45}$

$r = \frac{-30}{45}$

Reduce fractions to their simplest form (common factor of 15). $\underline{r = \frac{-2}{3}}$

91. Divide both sides of the equation by 0.2. $\frac{0.2c}{0.2} = \frac{5.8}{0.2}$

$c = \frac{5.8}{0.2}$

Divide. $c = 29$

92. Multiply both sides of the equation by 7. $7(\frac{x}{7}) = 7(16)$

$x = 7(16)$

Multiply. $\underline{x = 112}$

93. Multiply both sides of the equation by $^-4$. $^-4 \cdot \frac{y}{^-4} = {^-4} \cdot {^-12}$

$y = {^-4} \cdot {^-12}$

Signs the same? Multiply and give the result a positive sign. $\underline{y = {^+48} = 48}$

94. Divide both sides of the equation by $\frac{2}{3}$. $\frac{2}{3}a \div \frac{2}{3} = 54 \div \frac{2}{3}$

$a = 54 \div \frac{2}{3}$

Dividing by a fraction is the same as multiplying by its reciprocal. $a = 54 \times \frac{3}{2}$

Multiply the fractions on the right side of the equation. $a = \frac{54}{1} \times \frac{3}{2}$

$a = \frac{162}{2}$

$\underline{a = 81}$

95. Divide both sides of the equation by $\frac{8}{5}$. $\frac{8}{5}b \div \frac{8}{5} = -\frac{56}{15} \div \frac{8}{5}$

$b = -\frac{56}{15} \div \frac{8}{5}$

Dividing by a fraction is the same as multiplying by its reciprocal. $b = -\frac{56}{15} \times \frac{5}{8}$

Multiply the fractions on the right-side of the equation. Signs different?

Give the result a negative sign. $b = -\frac{280}{120}$

$\underline{b = -\frac{7}{3}}$

96. Let x = the suggested selling price of the car. The first and second sentences tell you that $\frac{7}{8}$ of the suggested price = \$28,000. So your equation is: $\frac{7}{8}x = \$28{,}000$

Divide both sides of the equation by $\frac{7}{8}$. $\frac{7}{8}x \div \frac{7}{8} = 28{,}000 \div \frac{7}{8}$

$x = 28{,}000 \div \frac{7}{8}$

Dividing by a fraction is the same as multiplying by its reciprocal. $x = 28{,}000 \cdot \frac{8}{7}$

$x = \$32{,}000$

<u>So, the suggested selling price of the can is \$32,000.</u>

97. Let b = the number of bears in each packing crate.
The first sentence tells you that the number of packing crates times the number of bears in each is equal to the total number of bears. Your equation is: $54b = 324$
Divide both sides of the equation by 54. $54b \div 54 = 324 \div 54$
$b = 324 \div 54$
Divide. $b = 6$
So, there are 6 bears in each crate.

98. Let t = the number of turtle hatchlings born.
The first sentence tells you that only 3% survive to adulthood. Three percent of the turtles born is 1,200. Your equation will be: $(3\%)t = 1{,}200$
The numerical equivalent of 3% is 0.03, so the equation becomes $0.03t = 1{,}200$.
Divide both sides of the equation by 0.03. $0.03t \div 0.03 = 1{,}200 \div 0.03$
$t = 1{,}200 \div 0.03$
Divide. $t = 40{,}000$
So, 40,000 turtle hatchlings are actually born.

99. Let c = the number of acres he planted last year.
1.5 times c is 300. $1.5c = 300$
Divide both sides of the equation by 1.5. $1.5c \div 1.5 = 300 \div 1.5$
$c = 300 \div 1.5$
Divide. $c = 200$
So, he planted 200 acres last year.

100. Let d = her annual salary. Five percent of her salary equals her yearly bonus. Your equation will be: $(5\%)d = \$6{,}000$
The numerical equivalent of 5% is 0.05, so the equation becomes $0.05d = 6{,}000$.
Divide both sides of the equation by 0.05. $0.05d \div 0.05 = 6{,}000 \div 0.05$
$d = 6{,}000 \div 0.05$
Divide. $d = \$120{,}000$
So, her annual salary is $120,000.

5

Solving Multi-Step Equations

Solving multi-step equations simply combines the work you have done in the previous chapters. The solution techniques for the two types of basic equations you worked on in Chapter 4 are both utilized in the equations in this chapter.

Tips for Solving Multi-Step Equations

- There are at least two ways to show multiplication. You may be used to seeing multiplication shown with an × like this: $5 \times 3 = 15$. In equations, this symbolism becomes confusing. In algebra, the convention is to show multiplication with either a · like this: $5 \cdot 3 = 15$, or with parentheses like this: $5(3) = 15$. Both conventions will be used in the answers, so you should try to get used to either one.
- Similarly, division can be shown using the standard division symbol ÷, as in $10 \div 2 = 5$. Or, it can be shown using a fraction bar like this: $10 \div 2 = \frac{10}{2} = 5$. Use and get used to both.
- Check your answers before looking at the answer solutions. Just substitute the value you find for the variable and work

each side of the equation as if it were a numerical expression. If the quantities you find are equal, then your solution is correct.

- Dividing by a fraction is the same as multiplying by its reciprocal. For example, $9 \div \frac{2}{3} = 9(\frac{3}{2})$.
- To write an equation for a word problem, let the unknown quantity be equal to the variable. Then, write the equation using the information stated in the problem.

For numbers 101–120, find the solutions to the given equation.

101. $4x + 7 = 11$

102. $^{-}15x + 11 = 86$

103. $3x - 8 = 16$

104. $5x - 6 = {}^{-}26$

105. $\frac{x}{3} + 4 = 10$

106. $\frac{^{-}x}{9} + \frac{1}{2} = \frac{11}{18}$

107. $39 = 3a - 9$

108. $4 = 4a + 20$

109. $^{-}11 = {}^{-}2a + 7$

110. $\frac{2}{3}m + 8 = 20$

111. $9 = \frac{3}{4}m - 3$

112. $\frac{2m}{5} + 16 = 24$

113. $^{-}1.2w - 7.5 = 8.1$

114. $^{-}5.12 = 0.4z - 1.88$

115. $0.3a + 0.25 = 1$

116. $7m - 6 = {}^{-}2.5$

117. $10s - 6 = 0$

118. $\frac{s}{4} + 2.7 = 3$

119. $\frac{^{-}2}{5}y + 0.1 = {}^{-}0.7$

120. $^{-}6.4z + \frac{4}{5} = \frac{^{-}12}{5}$

For numbers 121–125, solve the given word problem by letting a variable equal the unknown quantity, writing down an equation from the information given, and then solving the equation.

121. A farmer is raising a hog that weighed 20 lbs. when he bought it. He expects it to gain 12 pounds per month. He will sell it when it weighs 200 lbs. How many months will it be before he will sell the animal?

122. Mary earns $1.50 less than twice Bill's hourly wage. Mary earns $12.50 per hour. What is Bill's hourly wage?

123. At year's end, a share of stock in Axon Corporation was worth $37. This was $8 less than three times its value at the beginning of the year. What was the price of a share of Axon stock at the beginning of the year?

124. Jennifer earned $4,000 more than 1.5 times her former salary by changing jobs. She earned $64,000 at her new job. What was her salary at her previous employment?

125. Twenty-five more girls than $\frac{2}{3}$ the number of boys participate in interscholastic sports at a local high school. If the number of girls participating is 105, how many boys participate?

Answers

Numerical expressions in parentheses like this [ ] are operations performed on only part of the original expression. The operations performed within these symbols are intended to show how to evaluate the various terms that make up the entire expression.

Expressions with parentheses that look like this ( ) contain either numerical substitutions or expressions that are part of a numerical expression. Once a single number appears within these parentheses, the parentheses are no longer needed and need not be used the next time the entire expression is written.

When two pair of parentheses appear side by side like this ( )( ), it means that the expressions within are to be multiplied.

Sometimes parentheses appear within other parentheses in numerical or algebraic expressions. Regardless of what symbol is used, ( ), { }, or [ ], perform operations in the innermost parentheses first and work outward.

Underlined equations show the simplified result.

101. Subtract 7 from both sides of the equation. $4x + 7 - 7 = 11 - 7$
Associate like terms. $4x + (7 - 7) = (11 - 7)$
Perform numerical operations. $4x + (0) = (4)$
Zero is the identity element for addition. $4x = 4$
Divide both sides of the equation by 4. $\frac{4x}{4} = \frac{4}{4}$
$\underline{x = 1}$

102. Subtract 11 from both sides of the equation. $^-15x + 11 - 11 = 86 - 11$
Associate like terms. $^-15x + (11 - 11) = (86 - 11)$
Perform numerical operations. $^-15x + (0) = (75)$
Zero is the identity element of addition. $\underline{^-15x = 75}$
Divide both sides of the equation by $^-15$.
Signs different? Give the result a negative sign. $^-15x \div (^-15) = 75 \div (^-15)$
$\underline{x = {^-5}}$

103. Add 8 to each side of the equation. $3x - 8 + 8 = 16 + 8$
Change subtraction to addition and change the sign of the term that follows. $3x + {^-8} + 8 = 16 + 8$
Associate like terms. $3x + (^-8 + 8) = 16 + 8$
Perform numerical operations. $3x + (0) = 24$
Zero is the identity element for addition. $3x = 24$
Divide both sides of the equation by 3. $\frac{3x}{3} = \frac{24}{3}$
$\underline{x = 8}$

104. Add 6 to each side of the equation. $5x - 6 + 6 = {^-26} + 6$
Change subtraction to addition and change the sign of the term that follows. $5x + {^-6} + 6 = {^-26} + 6$
Associate like terms. $5x + (^-6 + 6) = (^-26 + 6)$
Perform numerical operations. $5x + (0) = {^-20}$
Zero is the identity element for addition. $5x = {^-20}$
Divide both sides of the equation by 5. $\frac{5x}{5} = \frac{-20}{5}$
$\underline{x = {^-4}}$

105. Subtract 4 from both sides of the equation. $\frac{x}{3} + 4 - 4 = 10 - 4$

Associate like terms. $\frac{x}{3} + (4 - 4) = (10 - 4)$

Perform numerical operations. $\frac{x}{3} + (0) = 6$

Zero is the identity element for addition. $\frac{x}{3} = 6$

Multiply both sides of the equation by 3. $3(\frac{x}{3}) = 3(6)$

$\underline{x = 18}$

106. Subtract $\frac{1}{2}$ from both sides of the equation. $-\frac{x}{9} + \frac{1}{2} - \frac{1}{2} = \frac{11}{18} - \frac{1}{2}$

Associate like terms. $-\frac{x}{9} + \left(\frac{1}{2} - \frac{1}{2}\right) = \left(\frac{11}{18} - \frac{1}{2}\right)$

Perform numerical operations. $\left[\frac{11}{18} - \frac{1}{2} = \frac{11}{18} - \frac{9}{18} = \frac{2}{18} = \frac{1}{9}\right]$

$-\frac{x}{9} + (0) = \left(\frac{1}{9}\right)$

Zero is the identity element of addition. $-\frac{x}{9} = \frac{1}{9}$

Multiply both sides of the equation by -9. $-\frac{x}{9} = \frac{1}{9}$

Signs different? Give the result a negative sign. $^-9\left(-\frac{x}{9}\right) = {}^-9\left(\frac{1}{9}\right)$

$\underline{x = {}^-1}$

107. Add 9 to each side of the equation. $39 + 9 = 3a - 9 + 9$

Change subtraction to addition and change the sign of the term that follows. $39 + 9 = 3a + ({}^-9 + 9)$

Associate like terms. $39 + 9 = 3a + ({}^-9 + 9)$

Perform numerical operations. $48 = 3a + (0)$

Zero is the identity element for addition. $48 = 3a$

Divide both sides of the equation by 3. $\frac{48}{3} = \frac{3a}{3}$

$\underline{16 = a}$

108. Subtract 20 from both sides of the equation. $4 - 20 = 4a + 20 - 20$

Associate like terms. $(4 - 20) = 4a + (20 - 20)$

Perform numerical operations. ${}^-16 = 4a + (0)$

Zero is the identity element for addition. ${}^-16 = 4a$

Divide both sides of the equation by 4. $\frac{^-16}{4} = \frac{4a}{4}$

$\underline{{}^-4 = a}$

109. Subtract 7 from both sides of the equation. ${}^-11 - 7 = {}^-2a + 7 - 7$

Associate like terms. $({}^-11 - 7) = {}^-2a + (7 - 7)$

Perform numerical operations. $({}^-18) = {}^-2a + (0)$

Zero is the identity element of addition. ${}^-18 = {}^-2a$

Divide both sides of the equation by $^-2$. $\frac{^-18}{^-2} = \frac{^-2a}{^-2}$

Signs the same? Give the result a positive sign. $9 = a$

$\underline{a = 9}$

110. Subtract 8 from both sides of the equation. $\frac{2}{3}m + 8 - 8 = 20 - 8$

Associate like terms. $\frac{2}{3}m + (8 - 8) = 20 - 8$

Perform numerical operations. $\frac{2}{3}m + (0) = 12$

Zero is the identity element for addition. $\frac{2}{3}m = 12$

Multiply both sides of the equation by the reciprocal of $\frac{2}{3}$. $\frac{3}{2}(\frac{2}{3}m) = \frac{3}{2}(12)$

$\underline{m = 18}$

111. Add 3 to both sides of the equation. $9 + 3 = \frac{3}{4}m - 3 + 3$

Change subtraction to addition and change the sign of the term that follows. $9 + 3 = \frac{3}{4}m + {}^{-}3 + 3$

Associate like terms. $(9 + 3) = \frac{3}{4}m + ({}^{-}3 + 3)$

Perform numerical operations. $12 = \frac{3}{4}m + (0)$

Zero is the identity element for addition. $12 = \frac{3}{4}m$

Multiply both sides of the equation by the reciprocal of $\frac{3}{4}$. $\frac{4}{3}(12) = \frac{4}{3}(\frac{3}{4}m)$

$\underline{16 = m}$

112. This equation presents a slightly different look. The variable in the numerator has a coefficient. There are two methods for solving.

Subtract 16 from both sides of the equation. $\frac{2m}{5} + 16 - 16 = 24 - 16$

Associate like terms. $\frac{2m}{5} + (16 - 16) = (24 - 16)$

Perform numerical operations. $\frac{2m}{5} + (0) = 8$

Zero is the identity element for addition. $\frac{2m}{5} = 8$

Multiply both sides of the equation by 5. $5(\frac{2m}{5}) = 5(8)$

Use rules for multiplying whole numbers and fractions. $\frac{5}{1}(\frac{2m}{5}) = 40$

$\frac{5 \cdot 2m}{1 \cdot 5} = 40$

$\frac{10m}{5} = 40$

$2m = 40$

Divide both sides by 2. $\frac{2m}{2} = \frac{40}{2}$

$m = 20$

Or you can recognize that $\frac{2m}{5} = (\frac{2}{5})m$. Then you would multiply by the reciprocal of the coefficient. $\frac{5}{2}(\frac{2}{5})m = (\frac{5}{2})8$

$\underline{m = 20}$

113. Add 7.5 to both sides of the equation. $^-1.2w - 7.5 + 7.5 = 8.1 + 7.5$

Change subtraction to addition and change the sign of the term that follows $^-1.2w + {}^-7.5 + 7.5 = 8.1 + 7.5$

Associate like terms. $^-1.2w + (^-7.5 + 7.5) = (8.1 + 7.5)$

Perform numerical operations. $^-1.2w + (0) = (15.6)$

Zero is the identity element of addition. $^-1.2w = 15.6$

Divide both sides of the equation by $^-1.2$. Signs different? $\frac{^-1.2w}{^-1.2} = \frac{15.6}{^-1.2}$

Give the result a negative sign. $\underline{w = {}^-13}$

114. Add 1.88 to both sides of the equation. $^-5.12 + 1.88 = 0.4z - 1.88 + 1.88$

Change subtraction to addition and change the sign of the term that follows. $^-5.12 + 1.88 = 0.4z + {}^-1.88 + 1.88$

Associate like terms. $(^-5.12 + 1.88) = 0.4z + (^-1.88 + 1.88)$

Perform numerical operations. $(^-3.24) = 0.4z + (0)$

Zero is the identity element of addition. $^-3.24 = 0.4z$

Divide both sides of the equation by 0.4. Signs different? $\frac{^-3.24}{0.4} = \frac{0.4z}{0.4}$

Give the result a negative sign. $^-8.1 = z$

$\underline{z = {}^-8.1}$

115. Subtract 0.25 from both sides of the equation. $0.3a + 0.25 - 0.25 = 1 - 0.25$

Associate like terms. $0.3a + (0.25 - 0.25) = 1 - 0.25$

Perform numerical operations. $0.3a + (0) = 0.75$

Zero is the identity element for addition. $0.3a = 0.75$

Divide both sides of the equation by 0.3. $\frac{0.3a}{0.3} = \frac{0.75}{0.3}$

Simplify the result. $\underline{a = 2.5}$

116. Add 6 to each side of the equation. $7m - 6 + 6 = {}^-2.5 + 6$

Change subtraction to addition and change the sign of the term that follows. $7m + {}^-6 + 6 = {}^-2.5 + 6$

Associate like terms. $7m + (^-6 + 6) = (^-2.5 + 6)$

Perform numerical operations. $7m + (0) = 3.5$

Divide both sides of the equation by 7. $\frac{7m}{7} = \frac{3.5}{7}$

$\underline{m = 0.5}$

117. Add 6 to each side of the equation. $10s - 6 + 6 = 0 + 6$

Change subtraction to addition and change the sign of the term that follows. $10s + {}^-6 + 6 = 0 + 6$

Associate like terms. $10s + ({}^-6 + 6) = (0 + 6)$

Perform numerical operations. $10s + (0) = 6$

Divide both sides of the equation by 10. $\frac{10s}{10} = \frac{6}{10}$

Express the answer in the simplest form. $\underline{s = \frac{3}{5} = 0.6}$

118. Subtract 2.7 from both sides of the equation. $\frac{s}{4} + 2.7 - 2.7 = 3 - 2.7$

Associate like terms. $\frac{s}{4} + (2.7 - 2.7) = (3 - 2.7)$

Perform numerical operations. $\frac{s}{4} + (0) = 0.3$

Multiply both sides of the equation by 4. $4(\frac{s}{4}) = 4(0.3)$

$\underline{s = 1.2}$

119. Subtract 0.1 from both sides of the equation. ${}^-\frac{2}{5}y + 0.1 - 0.1 = {}^-0.7 - 0.1$

Associate like terms. ${}^-\frac{2}{5}y + (0.1 - 0.1) = ({}^-0.7 - 0.1)$

Perform numerical operations. ${}^-\frac{2}{5}y + (0) = ({}^-0.8)$

Zero is the identity element of addition. ${}^-\frac{2}{5}y = {}^-0.8$

Multiply both sides of the equation by the reciprocal of ${}^-\frac{2}{5}$. Signs the same? $({}^-\frac{5}{2})({}^-\frac{2}{5}y) = ({}^-\frac{5}{2})({}^-0.8)$

$y = ({}^-\frac{5}{2})({}^-0.8) = \frac{5(0.8)}{2(1)}$

$y = \frac{4.0}{2} = 2$

Give the result a positive sign. $\underline{y = 2}$

120. Subtract $\frac{4}{5}$ from both sides of the equation. ${}^-6.4z + \frac{4}{5} - \frac{4}{5} = {}^-\frac{12}{5} - \frac{4}{5}$

Associate like terms. ${}^-6.4z + (\frac{4}{5} - \frac{4}{5}) = ({}^-\frac{12}{5} - \frac{4}{5})$

Perform numerical operations. ${}^-6.4z + (0) = ({}^-\frac{16}{5})$

Zero is the identity element of addition. ${}^-6.4z = {}^-\frac{16}{5}$

Divide both sides of the equation by ${}^-6.4$. Signs the same? Give the result a positive sign. ${}^-6.4z \div {}^-6.4 = {}^-\frac{16}{5} \div {}^-6.4$

$z = {}^-\frac{16}{5} \div -6.4$

$z = {}^-\frac{16}{5} \times {}^-\frac{1}{6.4} = \frac{16}{32}$

$\underline{z = \frac{1}{2}}$

121. Let x = the number of months. The number of months (x), times 12 (pounds per month), plus the starting weight (20), will be equal to 200 pounds. An equation that represents these words would be $12x + 20 = 200$.

Subtract 20 from both sides of the equation. $12x + 20 - 20 = 200 - 20$

Associate like terms. $12x + (20 - 20) = 200 - 20$

Perform numerical operations. $12x + (0) = 180$

Divide both sides of the equation by 12. $\frac{12x}{12} = \frac{180}{12}$

$x = 15$

The farmer would have to wait 15 months before selling his hog.

122. Let x = Bill's hourly wage. Then $2x$ less \$1.50 is equal to Mary's hourly wage. The equation representing the last statement would be $2x - 1.50 = 12.50$.

Add 1.50 to both sides of the equation. $2x - 1.50 + 1.50 = 12.50 + 1.50$

Perform numerical operations. $2x = 14.00$

Divide both sides of the equation by 2. $\frac{2x}{2} = \frac{14.00}{2}$

$x = 7.00$

Bill's hourly wage is \$7.00 per hour.

123. Let x = the share price at the beginning of the year. The statements tell us that if we multiply the share price at the beginning of the year by 3 and then subtract \$8, it will equal \$37. An equation that represents this amount is $3x - 8 = 37$.

Add 8 to both sides of the equation. $3x - 8 + 8 = 37 + 8$

Perform numerical operations. $3x = 45$

Divide both sides of the equation by 3. $\frac{3x}{3} = \frac{45}{3}$

$x = 15$

One share of Axon costs \$15 at the beginning of the year.

124. Let x = her former salary. The statements tell us that \$64,000 is equal to 1.5 times x plus \$4,000. An algebraic equation to represent this statement is $64{,}000 = 1.5x + 4{,}000$.

Subtract 4,000 from both sides of the equation. $64{,}000 - 4{,}000 = 1.5x + 4{,}000 - 4{,}000$

Perform numerical operations. $60{,}000 = 1.5x$

Divide both sides of the equation by 1.5. $\frac{60{,}000}{1.5} = \frac{1.5x}{1.5}$

$40{,}000 = x$

Jennifer's former salary was \$40,000 per year.

125. Let x = the number of boys who participate in interscholastic sports. The question tells us that $\frac{2}{3}$ the number of boys plus 25 is equal to the number of girls who participate. An equation that represents this statement is $\frac{2}{3}x + 25 = 105$.

Subtract 25 from both sides of the equation. $\frac{2}{3}x + 25 - 25 = 105 - 25$

Perform numerical operations. $\frac{2}{3}x = 80$

Multiply by the reciprocal of $\frac{2}{3}$. $\frac{3}{2}(\frac{2}{3}x) = \frac{3}{2}(80)$

$x = 120$

The number of boys who participate is 120.

6

Solving Equations with Variables on Both Sides of an Equation

If you have been solving the problems in this book successfully, you will move easily into this chapter. Work through the questions carefully, and refer to the answer explanations as you try and solve the equations by yourself. Then check your answers with the solutions provided. If your sequence of steps is not identical to the solution shown, but you are getting the correct answers, that's all right. There is often more than one way to find a solution. And it demonstrates your mastery of the processes involved in doing algebra.

Tips for Solving Equations with Variables on Both Sides of the Equation

Use the distributive property of multiplication to expand and separate terms. Notice that what follows are variations on the basic distributive property.

$a(b - c) = ab - ac$
$^-1(b + c) = {^-b} - c$
$^-1(b - c) = {^-b} + c$
$^-(b - c) = {^-b} + c$

The object is to isolate the variable on one side of the equation. When the variable stands alone on one side of the equation, you have found the solution. There are two instances that sometimes occur when solving equations. In the instance where you have eliminated the variable altogether from the equation and end up with two values that do not equal each other—such as 5 = 7, which we know is not true—there is said to be no solution for the equation. Stated another way, the solution is the null set—that is, a set containing no elements. In another instance, you find that the solution seems to be the variable, or some number, equal to itself. In that instance, any value can make the equation true; therefore, there are an infinite number of solutions.

As you begin to practice solving the following equations, remember that you can add, subtract, multiply, and divide variables on both sides of an equation just as you did with numerical values.

Find the solutions for the following equations.

126. $11x + 7 = 3x - 9$

127. $3x - 23 = 54 - 4x$

128. $5x + 3 + 6x = 10x + 9 - x$

129. $10x + 27 - 5x - 46 = 32 + 3x - 19$

130. $20x - 11 - 3x = 43 + 9x$

131. $0.4 + 3x - 0.25 = 1.15 - 2x$

132. $2x + 17 - 1.2x = 10 - 0.2x + 11$

133. $2 + 6x - 0.2 = 5x + 2.1$

134. $3x + 12 - 0.8x = 3.4 - 0.8x - 9.4$

135. $^-1.25 + 0.65x + 0.8 = {^-1.75} + 1.30x$

136. $4(4x + 3) = 6x - 28$

137. $7(x + 2) + 1 = 3(x + 14) - 4x$

138. $13 - 8(x - 2) = 7(x + 4) + 46$

139. $13x + 3(3 - x) = {^-3}(4 + 3x) - 2x$

140. $2(2x + 19) - 9x = 9(13 - x) + 21$

141. $12x - 4(x - 1) = 2(x - 2) + 16$

142. $\frac{5}{4}(8 - 2x) - \frac{1}{2}x = 22 + x$

143. $\frac{5}{2}(x - 2) + 3x = 3(x + 2) - 10$

144. $6(\frac{1}{2}x + \frac{1}{2}) = 3(x + 1)$

145. $0.7(0.2x - 1) = 0.3(3 - 0.2x)$

146. $10(x + 2) + 7(1 - x) = 3(x + 9)$

147. $4(9 - x) = 2x - 6(x + 6)$

148. $^{-}2[x - 2(3 + 4x)] = 2[1 - 2(1 - x)]$

149. $\frac{2}{3} - 2[\frac{1}{4}x + 2(1 - x)] = \frac{1}{12}(3 - 2x)$

150. $0.90 + 0.40[x - (2 - 0.50x)] = 1.1 - 1.40x$

Answers

Numerical expressions in parentheses like this [ ] are operations performed on only part of the original expression. The operations performed within these symbols are intended to show how to evaluate the various terms that make up the entire expression.

Expressions with parentheses that look like this ( ) contain either numerical substitutions or expressions that are part of a numerical expression. Once a single number appears within these parentheses, the parentheses are no longer needed and need not be used the next time the entire expression is written.

When two pair of parentheses appear side by side like this ( )( ), it means that the expressions within are to be multiplied.

Sometimes parentheses appear within other parentheses in numerical or algebraic expressions. Regardless of what symbol is used, ( ), { }, or [ ], perform operations in the innermost parentheses first and work outward.

Underlined equations show the simplified result.

126.	Subtract 7 from both sides of the equation.	$11x + 7 - 7 = 3x - 9 - 7$
	Simplify by combining like terms.	$11x + (0) = 3x - 16$
	Identity property of 0 for addition.	$11x = 3x - 16$
	Subtract $3x$ from both sides of the equation.	$11x - 3x = 3x - 3x - 16$
	Simplify.	$8x = {}^{-}16$
	Divide both sides of the equation by 8.	$\frac{8x}{8} = \frac{{}^{-}16}{8}$
	Simplify.	$1x = {}^{-}2$
	Solution.	$x = {}^{-}2$
127.	Add 23 to both sides of the equation.	$3x - 23 + 23 = 54 + 23 - 4x$
	Simplify by combining like terms.	$3x + 0 = 77 - 4x$
	Identity property of 0 for addition.	$3x = 77 - 4x$
	Now add $4x$ to both sides.	$3x + 4x = 77 - 4x + 4x$
	Simplify.	$7x = 77$
	Divide both sides of the equation by 7.	$\frac{7x}{7} = \frac{77}{7}$
	Simplify.	$x = 11$
128.	Use the commutative property of addition with like terms.	$5x + 6x + 3 = 10x - x + 9$
	Combine like terms on each side of the equation.	$11x + 3 = 9x + 9$
	Subtract 3 from both sides.	$11x + 3 - 3 = 9x + 9 - 3$
	Simplify by combining like terms.	$11x = 9x + 6$
	Now subtract $9x$ from both sides of the equation.	$11x - 9x = 9x - 9x + 6$
	Simplify.	$2x = 6$
	Divide both sides by 2.	$\frac{2x}{2} = \frac{6}{2}$
	Simplify.	$x = 3$

129. Use the commutative property of addition with like terms. $10x - 5x + 27 - 46 = 3x + 32 - 19$

Combine like terms on each side of the equation. $5x - 19 = 3x + 13$

Add 19 to both sides of the equation. $5x + 19 - 19 = 3x + 19 + 13$

Combine like terms on each side of the equation. $5x = 3x + 32$

Subtract $3x$ from both sides of the equation to isolate the variable on one side of the equation. $5x - 3x = 3x - 3x + 32$

Simplify. $2x = 32$

Divide both sides of the equation by 2. $\frac{2x}{2} = \frac{32}{2}$

Simplify. $\underline{x = 16}$

130. Use the commutative property to move like terms. $20x - 9x - 3x = 43 + 11$

Combine like terms on each side of the equation. $8x = 44$

Divide both sides of the equation by 4. $\frac{8x}{8} = \frac{54}{8}$

Simplify the expression. $\underline{x = 6.75}$

131. Use the commutative property to move like terms. $3x + 0.4 - 0.25 = 1.15 - 2x$

Combine like terms on each side of the equation. $3x + 0.15 = 1.15 - 2x$

Subtract 0.15 from both sides of the equation. $3x + 0.15 - 0.15 = 1.15 - 0.15 - 2x$

Combine like terms on each side of the equation. $3x + (0) = 1 - 2x$

Identity property of addition. $3x = 1 - 2x$

Add $2x$ to both sides of the equation. $3x + 2x = 1 - 2x + 2x$

Simplify. $5x = 1$

Divide both sides of the equation by 5. $\frac{5x}{5} = \frac{1}{5}$

Simplify the expression. $\underline{x = \frac{1}{5}}$

132. Use the commutative property with like terms. $2x - 1.2x + 17 = 10 + 11 - 0.2x$
Combine like terms on each side of the equation. $0.8x + 17 = 21 - 0.2x$
Subtract 17 from both sides of the equation. $0.8x + 17 - 17 = 21 - 17 - 0.2x$
Combine like terms on each side of the equation. $0.8x = 4 - 0.2x$
Add $0.2x$ to both sides of the equation. $0.8x + 0.2x = 4 - 0.2x + 0.2x$
Combine like terms on each side of the equation. $1x = 4$
Identity property of multiplication. $\underline{x = 4}$

133. Use the commutative property with like terms. $6x - 5x = 0.2 - 2 + 2.1$
Associate like terms on each side of the equation. $x = 0.3$
Simplify the expression. $x = 0.3$

134. Use the commutative property with like terms. $3x - 0.8x + 12 = 3.4 - 9.4 - 0.8x$
Associate like terms on each side of the equation. $(3x - 0.8x) + 12 = (3.4 - 9.4) - 0.8x$
Simplify the expression. $2.2x + 12 = {}^-6 - 0.8x$
Subtract 12 from both sides of the equation. $2.2x + 12 - 12 = {}^-6 - 12 - 0.8x$
Associate like terms on each side of the equation. $2.2x + (12 - 12) = ({}^-6 - 12) - 0.8x$
Simplify the expression. $2.2x + (0) = ({}^-18) - 0.8x$
Identity property of addition. $2.2x = {}^-18 - 0.8x$
Add $0.8x$ to both sides of the equation. $2.2x + 0.8x = {}^-18 + 0.8x - 0.8x$
Combine like terms on each side of the equation. $3x = {}^-18$
Divide both sides of the equation by 3. $\frac{3x}{3} = \frac{{}^-18}{3}$
Simplify the expression. $\underline{x = {}^-6}$

135. Associate like terms on each side of the equation. $(^{-}1.25 + 0.8) + 0.65x = ^{-}1.75 + 1.30x$

Simplify the expression. $(^{-}0.45) + 0.65x = ^{-}1.75 + 1.30x$

Remove the parentheses. $^{-}0.45 + 0.65x = ^{-}1.75 + 1.30x$

Add 0.45 to both sides of the equation. $^{-}0.45 + 0.45 + 0.65x = ^{-}1.75 + 0.45 + 1.30x$

Associate like terms on each side of the equation. $(^{-}0.45 + 0.45) + 0.65x = (^{-}1.75 + 0.45) + 1.30x$

Simplify the expression. $0.65x = ^{-}1.30 + 1.30x$

Subtract 1.30x from both sides of the equation. $0.65x - 1.30x = ^{-}1.30 + 1.30x - 1.30x$

Associate like terms on each side of the equation. $(0.65x - 1.30x) = ^{-}1.30 + (1.30x - 1.30x)$

Simplify the expression. $^{-}0.65x = ^{-}1.30$

Divide both sides of the equation by $^{-}0.65$. Signs the same? Give the result a positive sign. $\frac{^{-}0.65x}{^{-}0.65} = \frac{^{-}1.30}{^{-}0.65}$

$\underline{x = 2}$

136. Use the distributive property of multiplication. $4(4x) + 4(3) = 6x - 28$

Simplify the expression. $16x + 12 = 6x - 28$

Subtract 12 from both sides of the equation. $16x + 12 - 12 = 6x - 28 - 12$

Combine like terms on each side of the equation. $16x = 6x - 40$

Subtract $6x$ from both sides of the equation. $16x - 6x = 6x - 6x - 40$

Simplify the expression. $10x = ^{-}40$

Divide both sides of the equation by 10. $\frac{10x}{10} = \frac{^{-}40}{10}$

Simplify the expression. $\underline{x = ^{-}4}$

137.

Use the distributive property of multiplication.	$7(x) + 7(2) + 1 = 3(x) + 3(14) - 4x$
Simplify the expression.	$7x + 14 + 1 = 3x + 42 - 4x$
Use the commutative property with like terms.	$7x + 14 + 1 = 3x - 4x + 42$
Combine like terms on each side of the equation.	$7x + 15 = {}^{-}1x + 42$
Subtract 15 from both sides of the equation.	$7x + 15 - 15 = {}^{-}1x + 42 - 15$
Combine like terms on each side of the equation.	$7x = {}^{-}1x + 27$
Add x to both sides of the equation.	$7x + x = x + {}^{-}1x + 27$
Combine like terms on each side of the equation.	$8x = 27$
Divide both sides of the equation by 8.	$\frac{8x}{8} = \frac{27}{8}$
Simplify the expression.	$\underline{x = 3\frac{3}{8}}$

138.

Use the distributive property of multiplication.	$13 - 8(x) - 8({}^{-}2) = 7(x) + 7(4) + 46$
Simplify the expression.	$13 - 8x + 16 = 7x + 28 + 46$
Use the commutative property with like terms.	$13 + 16 - 8x = 7x + 28 + 46$
Combine like terms on each side of the equation.	$29 - 8x = 7x + 74$
Add $8x$ to both sides of the equation.	$29 + 8x - 8x = 8x + 7x + 74$
Combine like terms on each side of the equation.	$29 = 15x + 74$
Subtract 74 from both sides of the equation.	$29 - 74 = 15x + 74 - 74$
Combine like terms on each side of the equation.	${}^{-}45 = 15x$
Divide both sides of the equation by 15.	$\frac{{}^{-}45}{15} = \frac{15x}{15}$
Simplify the expression.	$\underline{{}^{-}3 = x}$

139. Use the distributive property of multiplication. $13x + 3(3) - 3(x) = {}^{-}3(4) - 3(3x) - 2x$

Simplify the expression. $13x + 9 - 3x = {}^{-}12 - 9x - 2x$

Use the commutative property with like terms. $9 + 13x - 3x = {}^{-}12 - 9x - 2x$

Combine like terms on each side of the equation. $9 + 10x = {}^{-}12 - 11x$

Add $11x$ to both sides of the equation. $9 + 10x + 11x = {}^{-}12 - 11x + 11x$

Combine like terms on each side of the equation. $9 + 21x = {}^{-}12$

Subtract 9 from both sides of the equation. $9 - 9 + 21x = {}^{-}12 - 9$

Combine like terms on each side of the equation. $21x = {}^{-}21$

Divide both sides of the equation by 21. $\frac{21x}{21} = \frac{{}^{-}21}{21}$

Simplify the expression. $\underline{x = {}^{-}1}$

140. Use the distributive property of multiplication. $2(2x) + 2(19) - 9x = 9(13) - 9(x) + 21$

Simplify the expression. $4x + 38 - 9x = 117 - 9x + 21$

Use the commutative property with like terms. $38 + 4x - 9x = 117 + 21 - 9x$

Combine like terms on each side of the equation. $38 - 5x = 138 - 9x$

Subtract 38 from both sides of the equation. $38 - 38 - 5x = 138 - 38 - 9x$

Combine like terms on each side of the equation. ${}^{-}5x = 100 - 9x$

Add $9x$ to both sides of the equation. ${}^{-}5x + 9x = 100 - 9x + 9x$

Combine like terms on each side of the equation. $4x = 100$

Divide both sides of the equation by 4. $\frac{4x}{4} = \frac{100}{4}$

Simplify the expression. $\underline{x = 25}$

141. Use the distributive property of multiplication. $12x - 4(x) - 4(^-1) = 2(x) + 2(^-2) + 16$
Simplify the expression. $12x - 4x + 4 = 2x - 4 + 16$
Combine like terms on each side of the equation. $8x + 4 = 2x + 12$
Subtract 4 from both sides of the equation. $8x + 4 - 4 = 2x + 12 - 4$
Simplify the expression. $8x = 2x + 8$
Subtract $2x$ from both sides of the equation. $8x - 2x = 2x - 2x + 8$
Simplify the expression. $6x = 8$
Divide both sides of the equation by 6. $\frac{6x}{6} = \frac{8}{6}$
Simplify the expression. $x = \frac{4}{3}$
Reduce fractions to simplest terms. $\underline{x = 1\frac{1}{3}}$

142. Use the distributive property of multiplication on the first term. $\frac{5}{4}(8) - \frac{5}{4}(2x) - \frac{1}{2}x = 22 + x$
$10 - \frac{5}{2}x - \frac{1}{2}x = 22 + x$
Associate like terms on each side of the equation. $10 + \left(^-\frac{5}{2}x - \frac{1}{2}x\right) = 22 + x$
Simplify the expression. $10 + (^-3x) = 22 + x$
Remove the parentheses. $10 - 3x = 22 + x$
Subtract 22 from both sides of the equation. $10 - 22 - 3x = 22 - 22 + x$
Associate like terms on each side of the equation. $(10 - 22) - 3x = (22 - 22) + x$
Simplify the expression. $^-12 - 3x = x$
Add to both sides of the equation. $^-12 - 3x + 3x = x + 3x$
Change subtraction to addition, and then associate like terms on each side of the equation. $^-12 + (^-3x + 3x) = (x + 3x)$
Simplify the expression. $^-12 = 4x$
Divide both sides of the equation by 4. Signs different? Give the result a negative sign. $\frac{^-12}{4} = \frac{4x}{4}$
$^-3 = x$
$\underline{x = ^-3}$

143. Use the distributive property on both sides. $\frac{5}{2}(x) - \frac{5}{2}(2) + 3x = 3(x) + 3(2) - 10$

Simplify the expression. $\frac{5}{2}x - 5 + 3x = 3x + 6 - 10$

Use the commutative property with like terms. $\frac{5}{2}x + 3x - 5 = 3x + 6 - 10$

A simple way to avoid having to operate with fractions is to multiply the equation by a factor that will eliminate the denominator in all fractions in the equation. In this case, that would be a 2. $2(\frac{5}{2}x + 3x - 5) = 2(3x + 6 - 10)$

Use the distributive property. $2(\frac{5}{2}x) + 2(3x) - 2(5) = 2(3x) + 2(6) - 2(10)$

Simplify the expressions. $5x + 6x - 10 = 6x + 12 - 20$

Combine like terms on each side of the equation. $11x - 10 = 6x - 8$

Add 10 to both sides of the equation. $11x - 10 + 10 = 6x - 8 + 10$

Simplify the expression. $11x = 6x + 2$

Subtract $6x$ from both sides of the equation. $11x - 6x = 6x - 6x + 2$

Combine like terms on each side of the equation. $5x = 2$

Divide both sides of the equation by 5. $\frac{5x}{5} = \frac{2}{5}$

Simplify the expression. $\underline{x = \frac{2}{5}}$

144. Use the distributive property on both sides. $6(\frac{1}{2}x) + 6(\frac{1}{2}) = 3(x) + 3(1)$

Simplify the expression. $3x + 3 = 3x + 3$

Subtract 3 from both sides of the equation. $3x + 3 - 3 = 3x + 3 - 3$

Simplify the expression. $3x = 3x$

Divide both sides of the equation by 3. $\frac{3x}{3} = \frac{3x}{3}$

Simplify the expression. $\underline{x = x}$

There are an infinite number of solutions for this equation because if any number is plugged in for x, you get a true statement.

145. Use the distributive property on both sides. $0.7(0.2x) - 0.7(1) = 0.3(3) - 0.3(0.2x)$
Simplify the expressions. $0.14x - 0.7 = 0.9 - 0.06x$
Add $0.06x$ to both sides of the equation. $0.14x + 0.06x - 0.7 = 0.9 - 0.06x + 0.06x$
Combine like terms on each side of the equation. $0.2x - 0.7 = 0.9$
Add 0.7 to both sides of the equation. $0.2x - 0.7 + 0.7 = 0.9 + 0.7$
Combine like terms on each side of the equation. $0.2x = 1.6$
Divide both sides of the equation by 0.2. $\frac{0.2x}{0.2} = \frac{1.6}{0.2}$
Simplify the expression. $\underline{x = 8}$

146. Use the distributive property of multiplication. $10(x) + 10(2) + 7(1) - 7(x) = 3(x) + 3(9)$
Simplify the expression. $10x + 20 + 7 - 7x = 3x + 27$
Use the commutative property with like terms. $10x - 7x + 20 + 7 = 3x + 27$
Combine like terms on each side of the equation. $3x + 27 = 3x + 27$
Look familiar? Subtract 27 from both sides. $3x + 27 - 27 = 3x + 27 - 27$
Simplify the expression. $3x = 3x$
Divide both sides of the equation by 3. $\underline{x = x}$

<u>There are an infinite number of solutions for this equation because if any number is plugged in for x, you get a true statement.</u>

147. Use the distributive property of multiplication. $4(9) - 4(x) = 2x - 6(x) - 6(6)$
Simplify the expressions. $36 - 4x = 2x - 6x - 36$
Combine like terms. $36 - 4x = {}^{-}4x - 36$
Add $4x$ to both sides of the equation. $36 + 4x - 4x = 4x - 4x - 36$
Combine like terms on each side of the equation. $36 = {}^{-}36$
This is a false statement. So, <u>there is no solution for this equation.</u> Another way of saying this is to say that the solution for this equation is the null set.

148. Apply the order of operations to simplify both sides of the equation.

$$^-2[x-2(3+4x)] = {^-2}[x-6-8x]$$
$$= {^-2}(x) - 2(^-6) - 2(^-8x)$$
$$= {^-2x} + 12 + 16$$
$$2[1-2(1-x)] = 2[1-2(1)-2(^-x)] = 2[1-2+2x]$$
$$= 2[^-1+2x]$$
$$= 2(^-1) + 2(2x)$$
$$= {^-2} + 4x$$

Substitute the expressions into the original equation $14x + 12 = {^-2} + 4x$

Subtract 12 from both sides of the equation. $14x + 12 - 12 = {^-2} - 12 + 4x$

Associate like terms on each side of the equation. $14x + (12 - 12) = (^-2 - 12) + 4x$

Simplify the expression. $14x = {^-14} + 4x$

Subtract $4x$ from both sides of the equation. $14x - 4x = {^-14} + 4x - 4x$

Associate like terms on each side of the equation. $(14x - 4x) = {^-14} + (4x - 4x)$

Simplify the expression. $10x = {^-14}$

Divide both sides of the equation by 10. Signs different? Give the result a negative sign.

$$\frac{10x}{10} = \frac{^-14}{10}$$
$$\underline{x = \frac{^-7}{5} = {^-1}\tfrac{2}{5}}$$

149. Apply the order of operations to simplify the left side of the equation.

$$^-2[\tfrac{1}{4}x + 2(1-x)] = {^-2}[\tfrac{1}{4}x + 2 - 2x]$$
$$= {^-2}[2 - \tfrac{7}{4}x]$$
$$= {^-2}(2) - 2(^-\tfrac{7}{4}x)$$
$$= {^-4} + \tfrac{7}{2}x$$

Substitute the expression into the original equation $\frac{2}{3} - 4 + \frac{7}{2}x = \frac{1}{12}(3 - 2x)$

Multiply both sides of the equation by 12 (to clear the fractions).

$12 \times \frac{2}{3} - 4 + \frac{7}{2}x = 12 \times \frac{1}{12}(3 - 2x)$

$12\frac{2}{3} + 12(^-4) + 12(\frac{7}{2}x) = 3 - 2x$

$8 - 48 + 42x = 3 - 2x$

Associate like terms on each side of the equation. $(8 - 48) + 42x = 3 - 2x$

Simplify the expression. $^-40 + 42x = 3 - 2x$

Add 40 to both sides of the equation. $40 + {^-40} + 42x = 40 + 3 - 2x$

Associate like terms on each side of the equation. $(40 + {^-40}) + 42x = (40 + 3) - 2x$

Simplify the expression. $42x = 43 - 2x$

Add $2x$ to both sides of the equation. $42x + 2x = 43 + 2x - 2x$

Associate like terms on each side of the equation.	$(42x + 2x) = 43 + (2x - 2x)$
Simplify the expression.	$44x = 43$
Divide both sides of the equation by 44. Signs the same? Give the result a positive sign.	$\frac{44x}{44} = \frac{43}{44}$ $\underline{x = \frac{43}{44}}$

150.

Apply the order of operations to simplify the left side of the equation.	$0.40[x - (2 - 0.50x)] = 0.40[x - 2 + 0.50x]$ $= 0.40[1.50x - 2]$ $= 0.40(1.50x) + 0.40(-2)$ $= 0.60x - 0.80$
Substitute the expression back into the equation.	$0.90 + 0.60x - 0.80 = 1.1 - 1.40x$
Multiply both sides of the equation by 10 (to clear the decimals).	$9 + 6x - 8 = 11 - 14x$
Use the commutative property to gather like terms together on the left side of the equation.	$9 - 8 + 6x = 11 - 14x$
Associate like terms on the left side of the equation.	$(9 - 8) + 6x = 11 - 14x$
Simplify the expression.	$(1) + 6x = 11 - 14x$
Remove the parentheses.	$1 + 6x = 11 - 14x$
Subtract 1 from both sides of the equation.	$1 - 1 + 6x = 11 - 1 - 14x$
Associate like terms on each side of the equation.	$(1 - 1) + 6x = (11 - 1) - 14x$
Simplify the expression.	$6x = 10 - 14x$
Add $14x$ to both sides of the equation.	$6x + 14x = 10 + 14x - 14x$
Associate like terms on each side of the equation.	$(6x + 14x) = 10 + (14x - 14x)$
Simplify the expression.	$20x = 10$
Divide both sides of the equation by 20. Signs the same? Give the result a positive sign.	$\frac{20x}{20} = \frac{10}{20}$ $\underline{x = \frac{1}{2}}$

7

Using Formulas to Solve Equations

Chances are you have been asked to use formulas to solve problems in math, science, social studies, or technology. Algebra is a useful subject to apply when faced with problems in these areas. In this chapter, you will have the chance to solve word problems that could pop up in any one of these subject areas. These word problems require you to find an unknown value in a formula. You will be using your algebra problem-solving skills in every problem.

Tips for Using Formulas to Solve Equations

Given a formula involving several variables, you will generally be given values for all but one. Then you will be asked to solve the equation for the missing variable. It can be helpful to list each variable with its given value. Put a question mark next to the equal sign in place of the value for the unknown variable.

Keep in mind the rules for order of operations.

Select from these formulas the appropriate one to solve the following word problems:

Volume of a rectangular solid: $V = lwh$ where l = length, w = width, h = height

Distance formula: $D = rt$ where r = rate, t = time

Simple interest: $I = prt$ where p = principal, r = interest rate, t = time in years
Area of a trapezoid: $A = \frac{1}{2}h(b_1 + b_2)$ where h = height and b_1 and b_2 are the bases
Fahrenheit/Celsius equivalence: $C = \frac{5}{9}(F - 32)$
Volume of a cylinder: $V = \pi r^2 h$ (use $\pi \approx 3.14$) where r = radius and h = height
Surface area of a cylinder: $S = 2\pi r(r + h)$ where r = radius of the base, h = height of cylinder (use $\pi \approx 3.14$)

151. Find the volume of a rectangular solid whose length is 12 cm, width is 8 cm, and height is 3 cm.

152. A rectangular container has a volume of 98 ft^3. If the length of the box is 7 ft, and its width is 3.5 ft, what is its height?

153. An airplane flies at an average velocity of 350 miles per hour. A flight from Miami to Aruba takes 3.5 hours. How far is it from Miami to Aruba?

154. A bicycle tour group planned to travel 68 miles between two New England towns. How long would the trip take if they averaged 17 miles per hour for the trip?

155. A hiking group wanted to travel along a chapter of the Appalachian Trail for 4 days. They planned to hike 6 hours per day and wanted to complete a trail chapter that was 60 miles long. What rate of speed would they have to average to complete the trail chapter as planned?

156. At an interest rate of 6%, how much interest would $12,000 earn over 2 years?

157. Over a three-year period, the total interest paid on a $4,500 loan was $1,620. What was the interest rate?

158. In order to earn $1,000 in interest over 2 years at an annual rate of 4%, how much principal must be put into a savings account?

159. How long will it take for a $3,000 savings account to double its value at a simple interest rate of 10%?

160. What is the area of a trapezoid whose height is 8 cm and whose bases are 14 cm and 18 cm?

161. What would be the height of a trapezoidal building if, at its base, it measured 80 feet, its roofline measured 40 feet, and the surface area of one side was 7,200 ft^2?

162. A trapezoid with an area of 240 square inches has a height of 12 inches. What are the lengths of the bases, if the lower base is three times the length of the upper?

163. You plan to visit a tropical island where the average daytime temperature is 40° Celsius. What would that temperature be on the Fahrenheit scale?

164. Jeff won't play golf if the temperature falls below 50° Fahrenheit. He is going to a country where the temperature is reported in Celsius. What would Jeff's low temperature limit be in that country? (Round your answer to the nearest degree.)

165. A steel drum has a base with a radius of 2 feet and a height of 4 feet. What is its volume in cubic feet?

166. Find the radius of a cylinder whose volume is 339.12 cubic centimeters and whose height is 12 centimeters.

167. The volume of a cylindrical aquarium tank is 13,565 cubic feet. If its radius is 12 feet, what is its height to the nearest foot?

168. The radius of the base of a cylinder is 20 cm. The height of the cylinder is 40 cm. What is its surface area?

169. What is the height of a cylinder whose surface area is 282.6 square inches and whose radius is 3 inches?

170. A cylinder has a surface area of 2,512 square feet. The height of the cylinder is three times the radius of the base of the cylinder. Find the radius and the height of the cylinder.

Answers

Numerical expressions in parentheses like this [ ] are operations performed on only part of the original expression. The operations performed within these symbols are intended to show how to evaluate the various terms that make up the entire expression.

Expressions with parentheses that look like this ( ) contain either numerical substitutions or expressions that are part of a numerical expression. Once a single number appears within these parentheses, the parentheses are no longer needed and need not be used the next time the entire expression is written.

When two pair of parentheses appear side by side like this ( )( ), it means that the expressions within are to be multiplied.

Sometimes parentheses appear within other parentheses in numerical or algebraic expressions. Regardless of what symbol is used, ( ), { }, or [ ], perform operations in the innermost parentheses first and work outward.

The simplified result is underlined.

151. Write the applicable formula. $V = lwh$
List the values for the variables. $V = ?$
$l = 12$ cm
$w = 8$ cm
$h = 3$ cm
Substitute the values into the formula. $V = (12)(8)(3)$
Simplify the expression. $V = 288$
Include the units. $\underline{V = 288 \text{ cm}^3}$

152. Write the appropriate formula. $V = lwh$
List the values for the variables. $V = 98$
$l = 7$
$w = 3.5$
$h = ?$
Substitute the values into the formula. $98 = (7)(3.5)h$
Simplify the expression. $98 = 24.5h$
Divide both sides of the equation by 24.5. $\frac{98}{24.5} = \frac{24.5h}{24.5}$
Simplify the expression. $4 = h$
Include the units. $\underline{h = 4 \text{ feet}}$

153. Write the applicable formula. $D = rt$
List the values for the variables. $D = ?$
$r = 350$
$t = 3.5$
Substitute the values for the variables. $D = (350)(3.5)$
Simplify the expression. $D = 1{,}225$
Include the units. $\underline{D = 1{,}225 \text{ miles}}$

154. Write the applicable formula. $D = rt$
List the values for the variables. $D = 68$
$r = 17$
$t = ?$
Substitute the values for the variables. $68 = (17)t$
Simplify the expression. $68 = 17t$
Divide both sides of the equation by 24. $\frac{68}{17} = \frac{17t}{17}$
Simplify the expression. $4 = t$
Include the units. $\underline{4 \text{ hours} = t}$

155. Write the applicable formula. $D = rt$
List the values for the variables. $D = 60$
$r = ?$
Calculate the total number of hours. $4 \text{ days} \cdot \frac{6 \text{ hrs}}{\text{day}} = 24 \text{ hrs}$
$t = 24$
Substitute the given values into the formula. $60 = r(24)$
Simplify the expression. $60 = 24r$
Divide both sides of the equation by 24. $\frac{60}{24} = \frac{24r}{24}$
Simplify the expression. $2.5 = r$
Include the units. $\underline{2.5 \text{ miles per hour} = r}$

156. Write the applicable formula. $I = prt$
List the values for the variables. $I = ?$
$p = 12{,}000$
$r = 6\% = 0.06$
$t = 2$
Substitute the given values into the formula. $I = (12{,}000)(0.06)(2)$
Simplify the expression. $I = 1{,}440$
Include the units. $\underline{I = \$1{,}440}$

157. Write the applicable formula. $I = prt$
List the values for the variables. $I = 1{,}620$
$p = 4{,}500$
$r = ?$
$t = 3$
Substitute the given values into the formula. $1{,}620 = (4{,}500)r(3)$
Simplify the expression. $1{,}620 = 13{,}500r$
Divide both sides of the equation by 13,500. $\frac{1{,}620}{13{,}500} = \frac{13{,}500r}{13{,}500}$
Simplify the expression. $0.12 = r$
Express as a percent. $\underline{12\% = r}$

158. Write the applicable formula. $I = prt$
List the values for the variables. $I = 1{,}000$
$p = ?$
$r = 4\% = 0.04$
$t = 2$

Substitute the given values into the formula.	$1{,}000 = p(0.04)(2)$
Simplify the expression.	$1{,}000 = 0.08p$
Divide both sides of the equation by 0.08.	$\frac{1{,}000}{0.08} = \frac{0.08p}{0.08}$
Simplify the expression.	$12{,}500 = p$
Include the units.	$\underline{\$12{,}500 = p}$

159.

Write the applicable formula.	$I = prt$
To double its value, the account would have to earn $3,000 in interest.	
List the values for the variables.	$I = 3{,}000$ $p = 3{,}000$ $r = 10\% = 0.10$ $t = ?$
Substitute the given values into the formula.	$3{,}000 = (3{,}000)(0.10)t$
Simplify the expression.	$3{,}000 = 300t$
Divide both sides of the equation by 300.	$\frac{3{,}000}{300} = \frac{300t}{300}$
Simplify the expression.	$10 = t$
Include the units.	$\underline{10 \text{ years} = t}$

160.

Write the applicable formula.	$A = \frac{1}{2}h(b_1 + b_2)$
List the values for the variables.	$A = ?$ $b_1 = 14$ $b_2 = 18$ $h = 8$
Substitute the given values into the formula.	$A = \frac{1}{2} \cdot 8 \cdot (14 + 18)$
Simplify the expression.	$A = \frac{1}{2} \cdot 8 \cdot 32$ $A = \frac{1}{2} \cdot 256$ $A = 128$
Include the units.	$\underline{A = 128 \text{ cm}^2}$

161.

Write the applicable formula.	$A = \frac{1}{2}h(b_1 + b_2)$
List the values for the variables.	$A = 7{,}200$ $b_1 = 40$ $b_2 = 80$ $h = ?$
Substitute the given values into the formula.	$7{,}200 = \frac{1}{2} \cdot h \cdot (40 + 80)$
Simplify the expression.	$7{,}200 = \frac{1}{2} \cdot h \cdot 120$ $7{,}200 = 60\,h$
Divide both sides of the equation by 60.	$\frac{7{,}200}{60} = \frac{60h}{60}$
Simplify the expression.	$120 = h$
Include the units.	$\underline{120 \text{ ft} = h}$

162. Write the applicable formula. $A = \frac{1}{2}h(b_1 + b_2)$
List the values for the variables.
$A = 240$
$b_1 = x$
$b_2 = 3x$
$h = 12$

Substitute the given values into the formula. $240 = \frac{1}{2} \cdot 12 \cdot (x + 3x)$

Simplify the expression. $240 = \frac{1}{2} \cdot 12 \cdot 4x$
$240 = 6 \cdot 4x$
$240 = 24x$

Divide both sides of the equation by 24. $\frac{240}{24} = \frac{24x}{24}$
Simplify the expression. $10 = x$
Substitute 10 for x in the list of variables. $\underline{b_1 = 10 \text{ in}}$
$b_2 = 3(10)$
$\underline{b_2 = 30 \text{ in}}$

163. Write the applicable formula. $C = \frac{5}{9}(F - 32)$
List the values for the variables.
$C = 40$
$F = ?$

Substitute the given values into the formula. $40 = \frac{5}{9}(F - 32)$
Multiply both sides of the equation by 9. $9(40) = 9[\frac{5}{9}(F - 32)]$
Simplify the expression. $360 = 5(F - 32)$
$360 = 5F - 5(32)$
$360 = 5F - 160$
Add 160 to both sides of the equation. $360 + 160 = 5F + 160 - 160$
Combine like terms on each side of the equation. $520 = 5F$
Divide both sides of the equation by 5. $\frac{520}{5} = \frac{5F}{5}$
Simplify the expression. $\underline{104^\circ = F}$

164. Write the applicable formula. $C = \frac{5}{9}(F - 32)$
List the values for the variables.
$C = ?$
$F = 50$

Substitute the given values into the formula. $C = \frac{5}{9}(50 - 32)$

Simplify the expression. $C = \frac{5}{9}(18)$
$C = \frac{5}{9}(\frac{18}{1})$
$\underline{C = 10^\circ}$

165. Write the applicable formula. $V = \pi r^2 h$
List the values for the variables. $V = ?$
$r = 2$
$h = 4$
$\pi \approx 3.14$
Substitute the given values into the formula. $V \approx (3.14)(2)^2(4)$
Simplify the expression. $V \approx 50.24$
Include the units. $\underline{V \approx 50.24 \text{ ft}^3}$

166. Write the applicable formula. $V = \pi r^2 h$
List the values for the variables. $V = 339.12$
$r = ?$
$h = 12$
$\pi \approx 3.14$
Substitute the given values into the formula. $339.12 \approx (3.14)r^2(12)$
Simplify the expression. $339.12 \approx 37.68\, r^2$
Divide both sides of the equation by 37.68. $\frac{339.12}{37.68} \approx \frac{37.68r^2}{37.68}$
Simplify the expression. $9 \approx r^2$
$3 \approx r$
Include the units. $\underline{3 \text{ cm} = r}$

167. Write the applicable formula. $V = \pi r^2 h$
List the values for the variables. $V = 13{,}565$
$r = 12$
$h = ?$
$\pi \approx 3.14$
Substitute the given values into the formula. $13{,}565 \approx (3.14)(12^2)h$
Simplify the expression. $13{,}565 \approx 452.16h$
Divide both sides of the equation by 452.16. $\frac{13{,}565}{452.16} \approx \frac{452.16h}{452.16}$
Simplify the expression. $30 \approx h$
Include the units. $\underline{30 \text{ ft} = h}$

168. Write the applicable formula. $S = 2\pi r(r + h)$
List the values for the variables. $S = ?$
$r = 20$
$h = 40$
$\pi \approx 3.14$
Substitute the given values into the formula. $S \approx 2(3.14)(20)(20 + 40)$
Simplify the expression. $S \approx 7{,}536$
Include the units. $\underline{S \approx 7{,}536 \text{ cm}^2}$

169. Write the applicable formula.	$S = 2\pi r(r + h)$
List the values for the variables.	$S = 282.6$
	$r = 3$
	$h = ?$
	$\pi \approx 3.14$
Substitute the given values into the formula.	$282.6 \approx 2(3.14)(3)(3 + h)$
Simplify the expression.	$282.6 \approx 18.84(3 + h)$
Use the distributive property of multiplication.	$282.6 \approx 18.84(3) + 18.84(h)$
Simplify the expression.	$282.6 \approx 56.52 + 18.84\,h$
Subtract 56.52 from both sides of the equation.	$282.6 - 56.52 = 56.52 - 56.52 + 18.84h$
Simplify the expression.	$226.08 \approx 18.84h$
Divide both sides of the equation by 18.84.	$\frac{226.08}{18.84} \approx \frac{18.84h}{18.84}$
Simplify the expression.	$12 \approx h$
Include the units.	$\underline{12 \text{ in} = h}$

170. Write the applicable formula.	$S = 2\pi r(r + h)$
List the values for the variables.	$S = 2{,}512$
	$r = x$
	$h = 3x$
	$\pi \approx 3.14$
Substitute the given values into the formula.	$2{,}512 \approx 2(3.14)(x)(x + 3x)$
Simplify the expression.	$2{,}512 \approx 6.28x(4x)$
Divide both sides of the equation by 25.12.	$\frac{2{,}512}{25.12} \approx \frac{25.12x^2}{25.12}$
Simplify the expression.	$100 \approx x^2$
Solve for x.	$10 \approx x$
Substitute the value for x into the values list.	$\underline{r = x = 10 \text{ ft}}$
	$\underline{h = 3x = 3(10) = 30 \text{ ft}}$

8

Graphing Linear Equations

This chapter asks you to graph linear equations. The graph of a linear equation is the set of ordered pairs that form a line on a coordinate axis. Every point on the line is a solution for the equation. One method for graphing the solution is to use a table with x and y values that constitute specific solutions for the equation. You select a value for x and solve for the y value. But in this chapter, we will focus a second method of graphing which involves the slope and y-intercept.

The slope and y-intercept method may require you to change an equation into the so-called slope-intercept form. That is, the equation with two variables must be written in the form $y = mx + b$. Written in this form, the m value is a number that represents the slope of the line and the b is a number that represents the y-intercept. The slope of a line is the ratio of the change in the y value over the change in the x value from one point on the line to another. From one point to another, the slope is the rise over the run. The y-intercept is the point where the graph of the line crosses the y-axis. Another way of saying that is: The y-intercept is the value of y that occurs when the value of x is 0.

Tips for Graphing Linear Equations

- Rewrite the given equation in the form $y = mx + b$.
- Use the b value to determine where the line crosses the y-axis. That is the point $(0,b)$.

- Use the value of m as the slope of the equation. Write the slope as a fraction. If the value of m is a whole number, the slope is the whole number over 1. The value of m is $= \frac{\text{change in } y}{\text{change in } x}$.
- If the value of m is negative, use a negative sign in only the numerator or the denominator, not both. For example, $-\frac{3}{4} = \frac{^-3}{4} = \frac{3}{^-4}$.

For numbers 171–195, graph the line using the slope and y-intercept method.

171. $y = 2x + 3$

172. $y = 5x - 2$

173. $y = {}^-2x + 9$

174. $y = \frac{3}{4}x - 1$

175. $y = \frac{5}{2}x - 3$

176. $y - 2x = 4$

177. $y + 3x = {}^-2$

178. $2x + 5y = 30$

179. $x - 3y = 12$

180. ${}^-5x - y = -\frac{7}{2}$

181. $x = 7y - 14$

182. $0 = 3x + 2y$

183. $y - 0.6x = {}^-2$

184. $\frac{2}{3}y - \frac{1}{2}x = 0$

185. $\frac{5}{6}x - \frac{1}{3}y = 2$

186. $20x - 15 = 5y$

187. $0.1x = 0.7y + 1.4$

188. ${}^-34x + 85 = 17y$

189. $2y - 3x = 6 + 3x$

190. $1 - ({}^-2x + 3) = y + 3x$

191. ${}^-(3x + y) = 2(y - 2x)$

192. $1 - x - 3y = {}^-2(3y + 2x - 1)$

193. $(\frac{2}{3}x - \frac{1}{9}y) - (2 - \frac{4}{3}x) = \frac{2}{9}y + 1$

194. $\frac{1}{5}({}^-y + 10x) = y - x + 2$

195. $(0.1 - 0.5y - 0.4x) + (0.4y + 0.3x + 0.4) = 0$

For numbers 196–200, use the slope/*y*-intercept method to write an equation that would enable you to draw a graphic solution for each problem.

196. A glider has a 25:1 descent ratio when there are no updrafts to raise its altitude. That is, for every 25 feet it moves parallel to the ground, it will lose 1 foot of altitude. Write an equation to represent the glider's descent from an altitude of 250 feet.

197. An Internet service provider charges $15 plus $0.25 per hour of usage per month. Write an equation that would represent the monthly bill of a user.

198. A scooter rental agency charges $20 per day plus $0.05 per mile for the rental of a motor scooter. Write an equation to represent the cost of one day's rental.

199. A dive resort rents scuba equipment at a weekly rate of $150 per week and charges $8 per tank of compressed air used during the week of diving. Write an equation to represent a diver's cost for one week of diving at the resort.

200. A recent backyard bird count showed that one out of every seven birds that visited backyard feeders was a chickadee. Write an equation to represent this ratio.

Answers

Numerical expressions in parentheses like this [ ] are operations performed on only part of the original expression. The operations performed within these symbols are intended to show how to evaluate the various terms that make up the entire expression.

Expressions with parentheses that look like this ( ) contain either numerical substitutions or expressions that are part of a numerical expression. Once a single number appears within these parentheses, the parentheses are no longer needed and need not be used the next time the entire expression is written.

When two pair of parentheses appear side by side like this ( )( ), it means that the expressions within are to be multiplied.

Sometimes parentheses appear within other parentheses in numerical or algebraic expressions. Regardless of what symbol is used, ( ), { }, or [ ], perform operations in the innermost parentheses first and work outward.

Underlined expressions show the original algebraic expression as an equation with the expression equal to its simplified result.

171. The equation is already in the proper slope/y-intercept form.

slope: $m = \frac{\text{change in } y}{\text{change in } x} = 2 = \frac{2}{1}$

y-intercept: $b = 3$, so the y-intercept is the point (0,3).
Point on the line: Starting at (0,3), a change in y of 2 and in x of 1 gives the point (0+1, 3+2) = (1, 5).

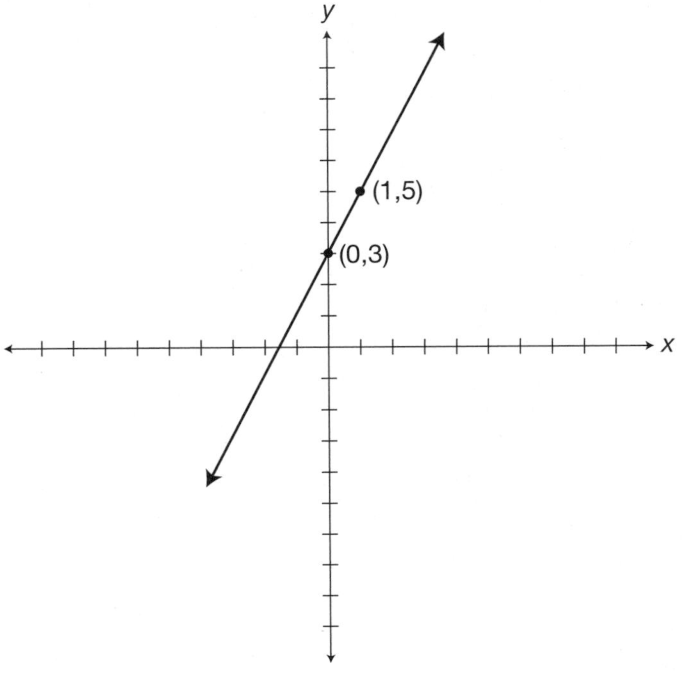

172. The equation is already in the proper slope/y-intercept form.

slope: $m = \dfrac{\text{change in } y}{\text{change in } x} = 5 = \frac{5}{1}$

y-intercept: $b = {}^{-}2$, so the y-intercept is the point $(0,{}^{-}2)$.

Point on the line: Starting at the point $(0,{}^{-}2)$, a change in y of 5 and in x of 1 gives the point $(0 + 1,{}^{-}2 + 5)$ or $(1,3)$.

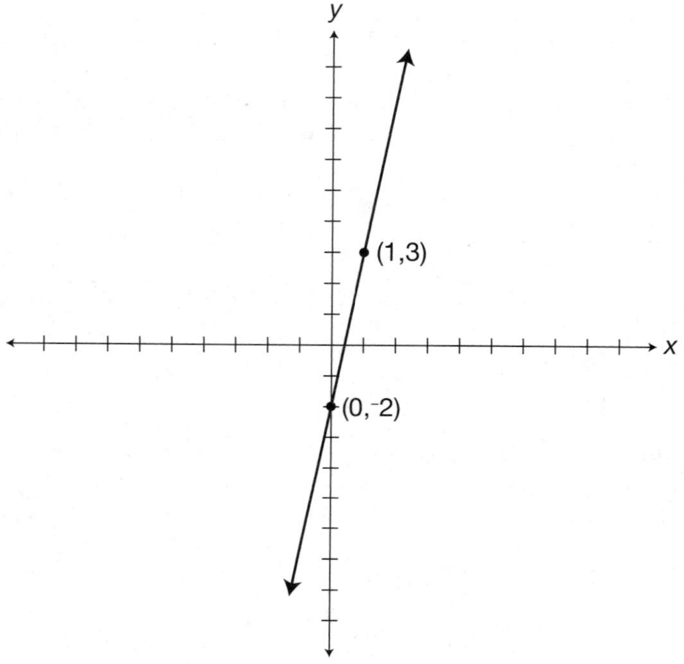

173. The equation is already in the proper slope/y-intercept form.

slope: $m = \frac{\text{change in } y}{\text{change in } x} = {}^{-}2 = {}^{-}\frac{2}{1}$

y-intercept: $b = 9$, so the y-intercept is the point (0,9).

Point on the line: Starting at the point (0,9), a change in y of ${}^{-}2$ and in x of 1 gives the point$(0 + 1,9 - 2)$ or (1,7).

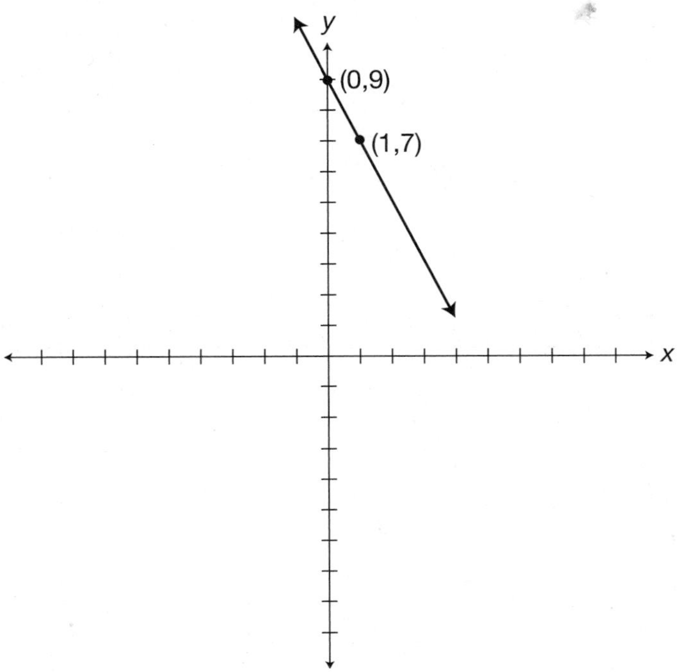

174. The equation is already in the proper slope/y-intercept form.

slope: $m = \frac{\text{change in } y}{\text{change in } x} = \frac{3}{4}$

y-intercept: $b = {}^-1$, so the y-intercept is the point $(0,{}^-1)$.

Point on the line: Starting at the point $(0,{}^-1)$, a change in y of 3 and in x of 4 gives the point $(0 + 4,{}^-1 + 3)$ or $(4,2)$.

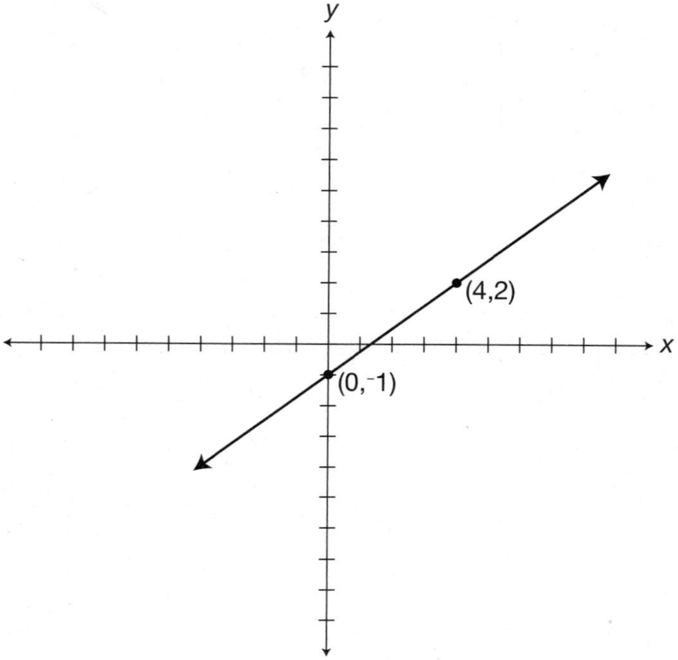

175. The equation is already in the proper slope/y-intercept form.

slope: $m = \frac{\text{change in } y}{\text{change in } x} = \frac{5}{2}$

y-intercept: $b = {}^-3$, so The y-intercept is the point $(0,{}^-3)$.

Point on the line: Starting at the point $(0,{}^-3)$, a change in y of 5 and in x of 2 gives the point$(0 + 2,{}^-3 + 5)$ or $(2,2)$.

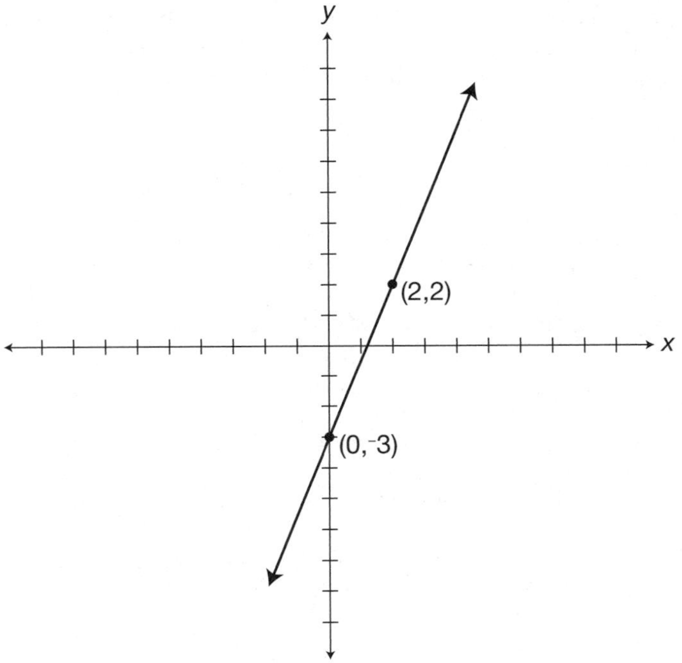

176. Put the equation in the proper form.
Add $2x$ to both sides of the equation. $y + 2x - 2x = 2x + 4$
Simplify the equation. $y = 2x + 4$
The equation is in the proper slope/y-intercept form.

<u>slope:</u> $m = \frac{\text{change in } y}{\text{change in } x} = 2 = \frac{2}{1}$

<u>y-intercept:</u> $b = 4$, so the y-intercept is the point (0,4).

<u>Point on the line:</u> Starting at the point (0,4), a change in y of 2 and in x of 1 gives the point (0 + 1,4 + 2) or (1,6).

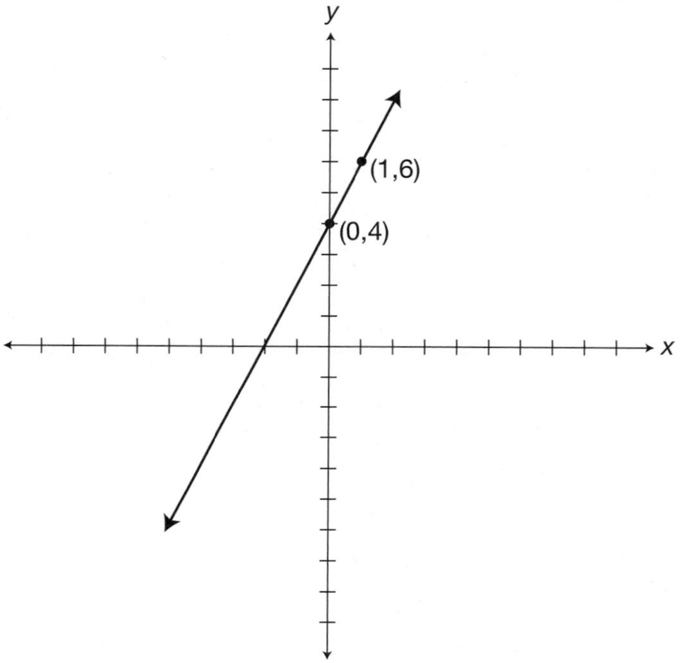

177. Put the equation in the proper form.

Subtract $3x$ from both sides of the equation. $y + 3x - 3x = {}^-3x - 2$

Simplify the equation. $y = {}^-3x - 2$

The equation is in the proper slope/y-intercept form.

<u>slope:</u> $m = \frac{\text{change in } y}{\text{change in } x} = {}^-3 = \frac{{}^-3}{1}$

<u>y-intercept:</u> $b = {}^-2$, so the y-intercept is the point $(0,{}^-2)$.

<u>Point on the line:</u> Starting at the point $(0,{}^-2)$, a change in y of $^-3$ and in x of 1 gives the point$(0 + 1,{}^-2 - 3)$ or $(1,{}^-5)$.

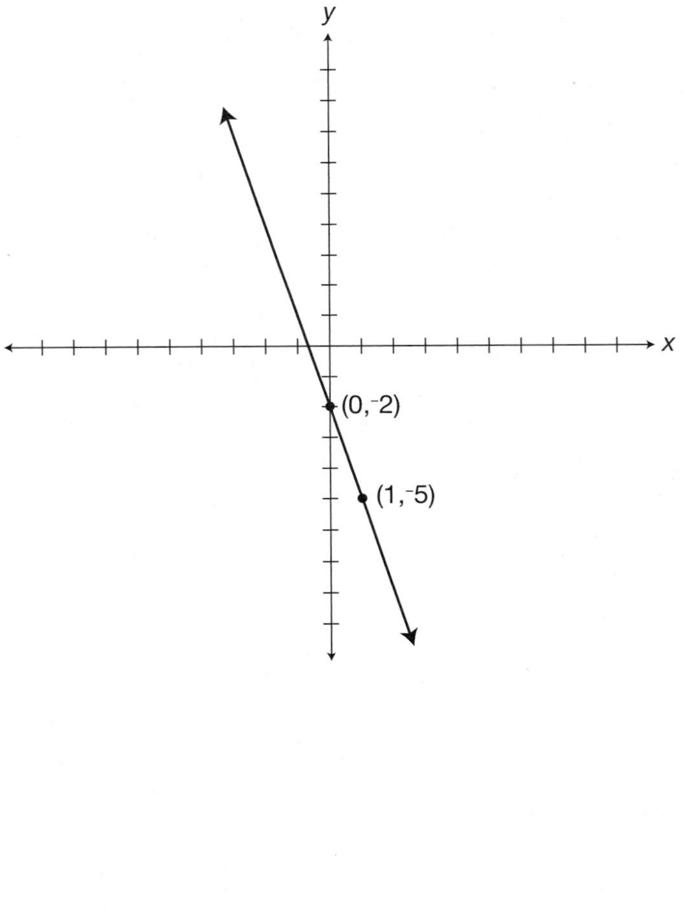

178. Put the equation in the proper form.

Subtract $2x$ from both sides of the equation. $2x - 2x + 5y = {}^-2x + 30$

Simplify the equation. $5y = {}^-2x + 30$

Divide both sides of the equation by 5. $\frac{5y}{5} = \frac{({}^-2x + 30)}{5}$

Simplify the equation. $y = \frac{{}^-2x}{5} + \frac{30}{5}$

$$y = \frac{{}^-2x}{5} + 6$$

The equation is now in the proper slope/y-intercept form.

<u>slope:</u> $m = \frac{\text{change in } y}{\text{change in } x} = \frac{-2}{5}$

<u>y-intercept:</u> $b = 6$. The y-intercept is the point (0,6).

<u>Point on the line:</u> Starting at the point (0,6), a change in y of $^-2$ and in x of 5 gives the point $(0 + 5, 6 - 2)$ or (5,4).

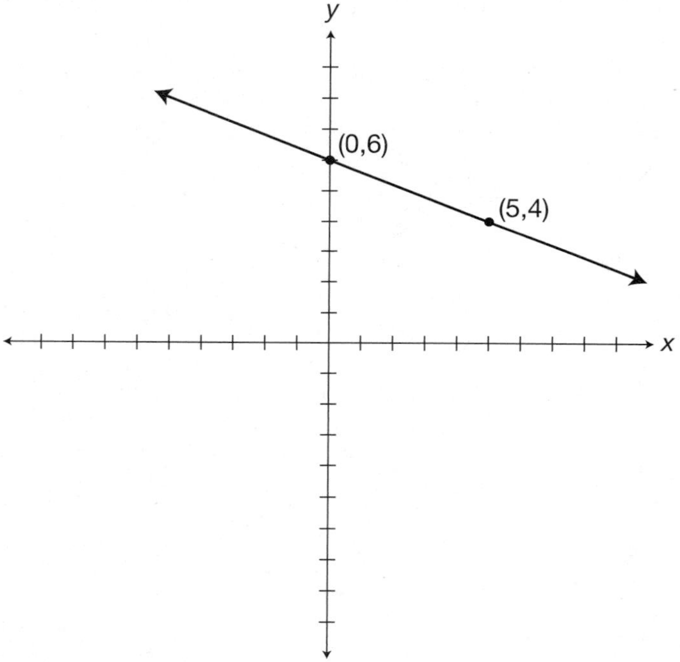

179. Put the equation in the proper form.

Add $3y$ to both sides of the equation.	$x - 3y + 3y = 12 + 3y$
Simplify the equation.	$x = 12 + 3y$
Subtract 12 from both sides of the equation.	$x - 12 = 12 - 12 + 3y$
Simplify the equation.	$x - 12 = 3y$
Divide both sides of the equation by 3.	$\frac{(x-12)}{3} = y$
Simplify the equation.	$\frac{x}{3} - \frac{12}{3} = y$
	$\frac{x}{3} - 4 = y$
	$\frac{x}{3} = \frac{(1)(x)}{(3)(1)} = \frac{1}{3} \cdot \frac{x}{1} = \frac{1}{3}x$
	$\frac{1}{3}x - 4 = y$
The equation is equivalent to the proper form.	$y = \frac{1}{3}x - 4$

The equation is now in the proper slope/y-intercept form.

<u>slope:</u> $m = \frac{\text{change in } y}{\text{change in } x} = \frac{1}{3}$

<u>y-intercept:</u> $b = {}^-4$. The y-intercept is the point $(0,{}^-4)$.

<u>Point on the line:</u> Starting at the point $(0,{}^-4)$, a change in y of 1 and in x of 3 gives the point $(0 + 3,{}^-4 + 1)$ or $(3,{}^-3)$.

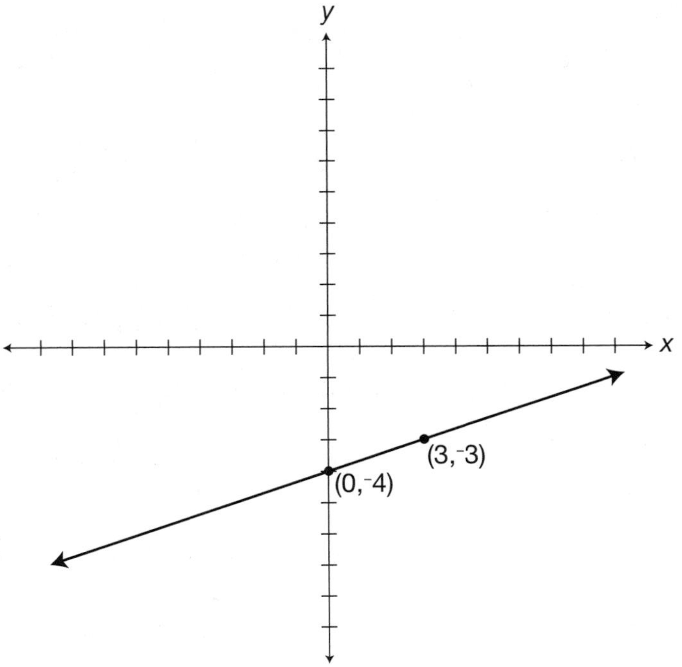

180. Put the equation in the proper form.

Add $5x$ to both sides of the equation. $5x - 5x - y = 5x - \frac{7}{2}$

Simplify the equation. $^-y = 5x - \frac{7}{2}$

Multiply both sides of the equation by $^-1$. $^-1(^-y) = {}^-1(5x - \frac{7}{2})$

Simplify the equation. $y = {}^-5x + \frac{7}{2}$

The equation is now in the proper slope/y-intercept form.

<u>slope:</u> $m = \frac{\text{change in } y}{\text{change in } x} = {}^-5 = {}^-\frac{5}{1}$

<u>y-intercept:</u> $b = \frac{7}{2} = 3\frac{1}{2}$. The y-intercept is the point $(0,3\frac{1}{2})$.

<u>Point on the line:</u> Starting at the point $(0,3\frac{1}{2})$, a change in y of $^-5$ and in x of 1 gives the point of $(0 + 1,3\frac{1}{2} - 5)$ or $(1,{}^-1\frac{1}{2})$.

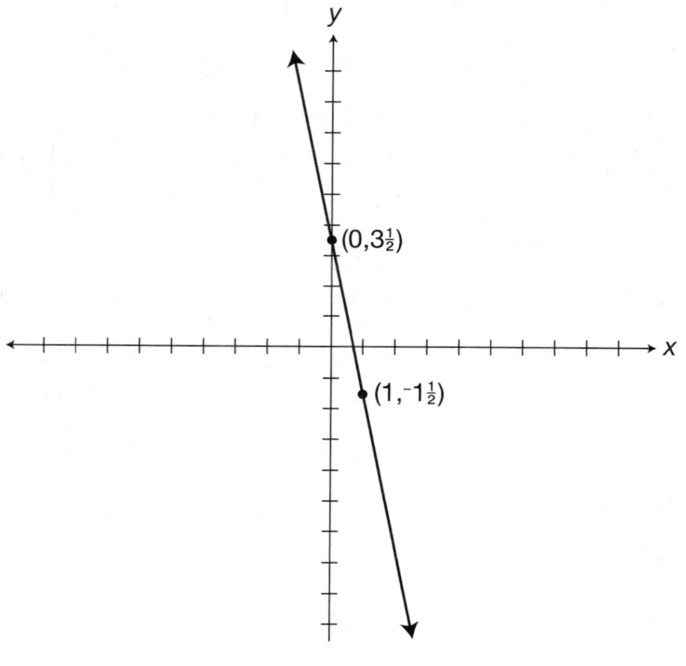

181. Put the equation in the proper form.

Add 14 to both sides of the equation. $x + 14 = 7y + 14 - 14$

Simplify the equation. $x + 14 = 7y$

Divide both sides of the equation by 7. $\frac{(x + 14)}{7} = \frac{7y}{7}$

Simplify the equation. $\frac{x}{7} + 2 = y$

$\frac{1}{7}x + 2 = y$

$y = \frac{1}{7}x + 2$

The equation is now in the proper slope/y-intercept form.

slope: $m = \frac{\text{change in } y}{\text{change in } x} = \frac{1}{7}$

y-intercept: $b = 2$. The y-intercept is the point (0,2).

Point on the line: Starting at the point (0,2), a change in y of 1 and in x of 7 gives the point (0 + 7,2 + 1) or (7,3).

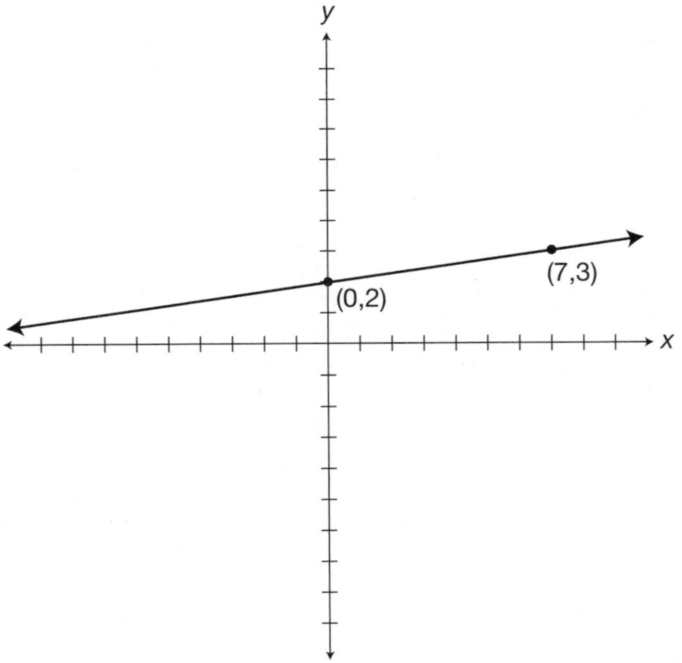

182. Put the equation in the proper form.
Subtract $2y$ from both sides of the equation. $0 - 2y = 3x + 2y - 2y$
Simplify the equation. $^-2y = 3x$
Divide both sides of the equation by $^-2$. $\frac{^-2y}{^-2} = \frac{3x}{^-2}$
Simplify the equation. $y = \frac{^-3}{2}x$
The equation is now in the proper slope/y-intercept form.

<u>slope:</u> $m = \frac{\text{change in } y}{\text{change in } x} = \frac{^-3}{2}$

There is no b showing in the equation, so $b = 0$. The y-intercept is the point (0,0).

<u>Point on the line:</u> Starting at the point (0,0), a change in y of $^-3$ and in x of 2 gives the point $(0 + 2, 0 - 3)$ or $(2, ^-3)$.

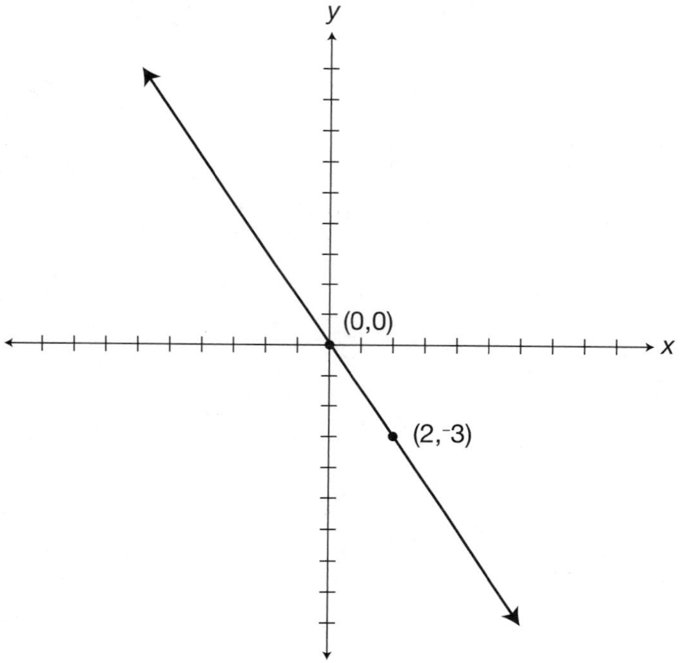

183. Put the equation in the proper form.
Add $0.6x$ to both sides of the equation. $y + 0.6x - 0.6x = 0.6x - 2$
Simplify the equation. $y = 0.6x - 2$
The equation is now in the proper slope/y-intercept form.

slope: $m = \frac{\text{change in } y}{\text{change in } x} = 0.6 = \frac{6}{10} = \frac{3}{5}$

y-intercept: $b = {}^-2$. The y-intercept is the point $(0,{}^-2)$.

Point on the line: Starting at the point $(0,{}^-2)$, a change in y of 3 and in x of 5 gives the point $(0 + 5,{}^-2 + 3)$ or $(5,1)$.

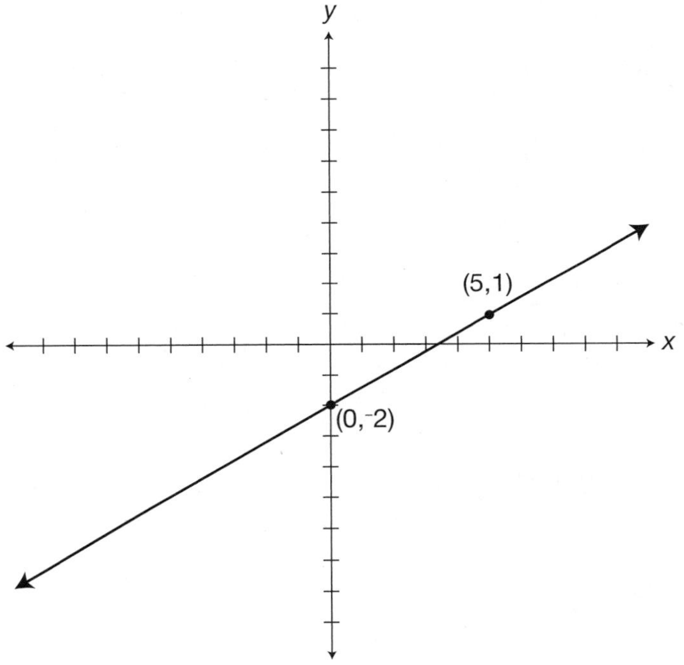

184. Put the equation in the proper form.

Add $\frac{1}{2}x$ to both sides of the equation. $\frac{2}{3}y - \frac{1}{2}x + \frac{1}{2}x = 0 + \frac{1}{2}x$

Simplify the equation. $\frac{2}{3}y = \frac{1}{2}x$

Multiply both sides of the equation by $\frac{3}{2}$. $\frac{3}{2}(\frac{2}{3})y = \frac{3}{2}(\frac{1}{2}x)$

Simplify the equation. $1y = \frac{3}{4}x$

$y = \frac{3}{4}x$

The equation is now in the proper slope/y-intercept form.

<u>slope:</u> $m = \frac{\text{change in } y}{\text{change in } x} = \frac{3}{4}$

<u>y-intercept:</u> $b = 0$. The y-intercept is the point (0,0).

<u>Point on the line:</u> Starting at the point (0,0), a change in y of 3 and in x of 4 gives the point (0 + 4,0 + 3) or (4,3).

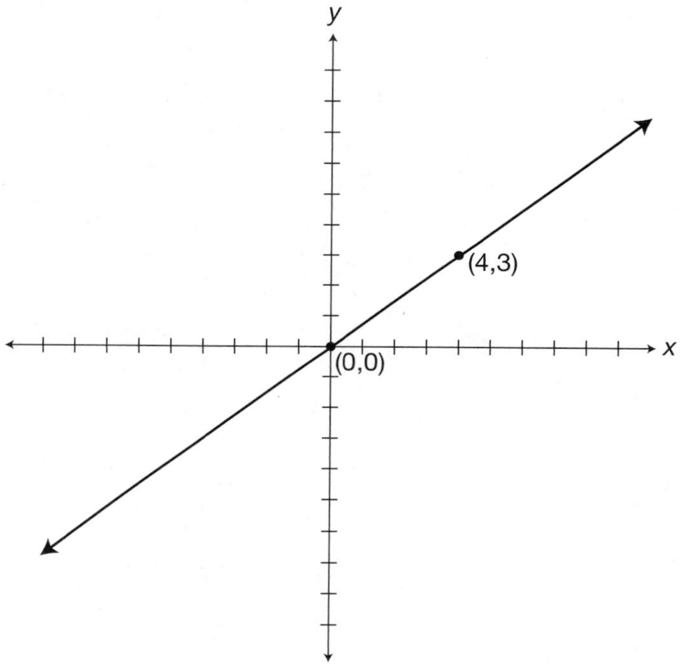

185. Simplify the equation. It would be easier to operate with an equation that doesn't have fractional coefficients. So, if you multiply the whole equation by the lowest common multiple of the denominators, you will have whole numbers with coefficients.

Multiply both sides of the equation by 6.	$6(\frac{5}{6}x - \frac{1}{3}y) = 6(2)$
Use the distributive property of multiplication.	$6(\frac{5}{6}x) - 6(\frac{1}{3}y) = 6(2)$
Simplify the equation.	$5x - 2y = 12$
Subtract $5x$ from both sides of the equation.	$5x - 5x - 2y = {}^-5x + 12$
Simplify the equation.	${}^-2y = {}^-5x + 12$
Divide both sides of the equation by $^-2$.	$\frac{^-2y}{^-2} = \frac{^-5x}{^-2} + \frac{12}{^-2}$
Simplify the equation.	$y = \frac{5}{2}x - 6$

The equation is now in the proper slope/y-intercept form.

slope: $m = \frac{\text{change in } y}{\text{change in } x} = \frac{5}{2}$

y-intercept: $b = {}^-6$. The y-intercept is the point $(0,{}^-6)$.

Point on the line: Starting at the point $(0,{}^-6)$, a change in y of 5 and in x of 2 gives the point $(0 + 2,{}^-6 + 5)$ or $(2,{}^-1)$.

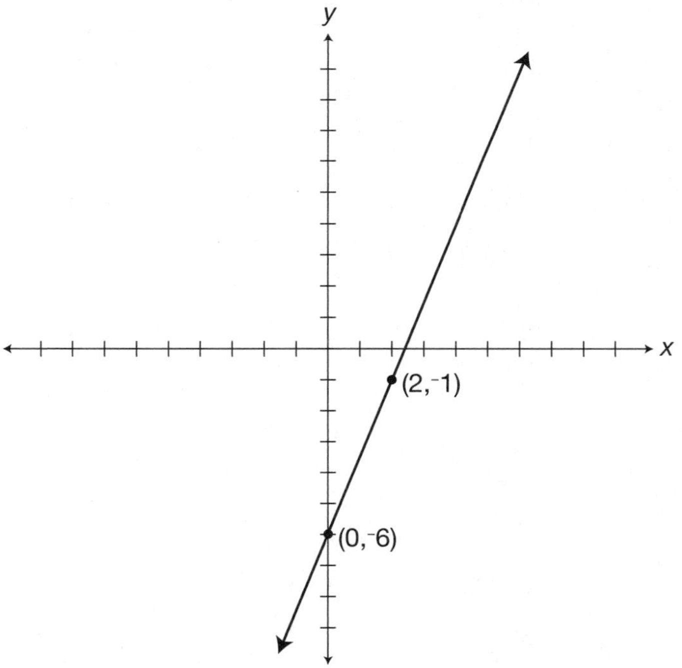

186. Exchange the terms on each side of the equal sign. $5y = 20x - 15$
Divide both sides of the equation by 5. $\frac{5y}{5} = \frac{20x}{5} - \frac{15}{5}$
Simplify the equation. $y = 4x - 3$
The equation is now in the proper slope/*y*-intercept form.

<u>slope:</u> $m = \frac{\text{change in } y}{\text{change in } x} = 4 = \frac{4}{1}$

<u>*y*-intercept:</u> $b = {}^-3$. The *y*-intercept is the point $(0,{}^-3)$.

<u>Point on the line:</u> Starting at the point $(0,{}^-3)$, a change in y of 4 and in x of 1 gives the point $(0 + 1,{}^-3 + 4)$ or $(1,1)$.

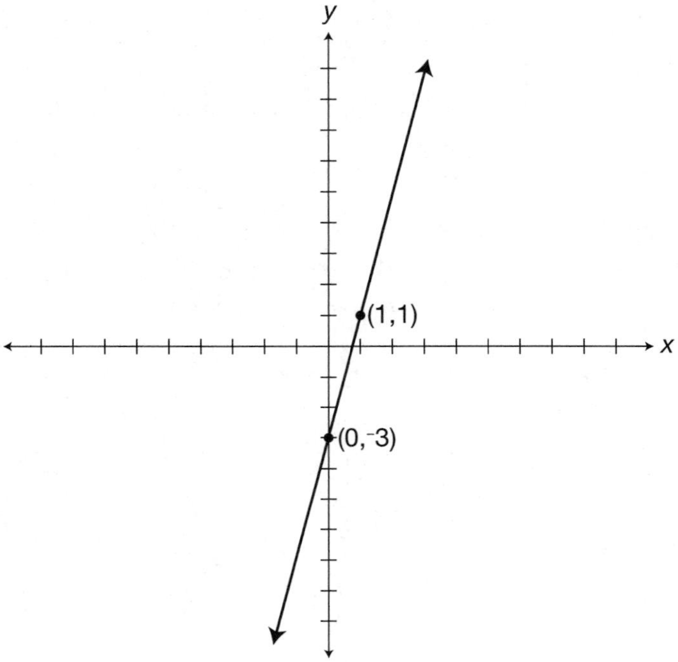

187. Once again, it is easier to operate with whole number coefficients instead of decimals to start. So, multiply the whole equation by 10.

Multiply both sides of the equation by 10.	$10(0.1x) = 10(0.7y + 1.4)$
Simplify the expression.	$x = 7y + 14$
Subtract 14 from both sides of the equation.	$x - 14 = 7y + 14 - 14$
Simplify the equation.	$x - 14 = 7y$
If $a = b$, then $b = a$.	$7y = x - 14$
Divide both sides of the equation by 7.	$\frac{7y}{7} = \frac{x}{7} - \frac{14}{7}$
Simplify the equation.	$y = \frac{1}{7}x - 2$

The equation is now in the proper slope/y-intercept form.

<u>slope:</u> $m = \frac{\text{change in } y}{\text{change in } x} = \frac{1}{7}$

<u>y-intercept:</u> $b = ^-2$. The y-intercept is the point $(0,^-2)$.

<u>Point on the line:</u> Starting at the point $(0,^-2)$, a change in y of 1 and in x of 7 gives the point $(0 + 7,^-2 + 1)$ or $(7,^-1)$.

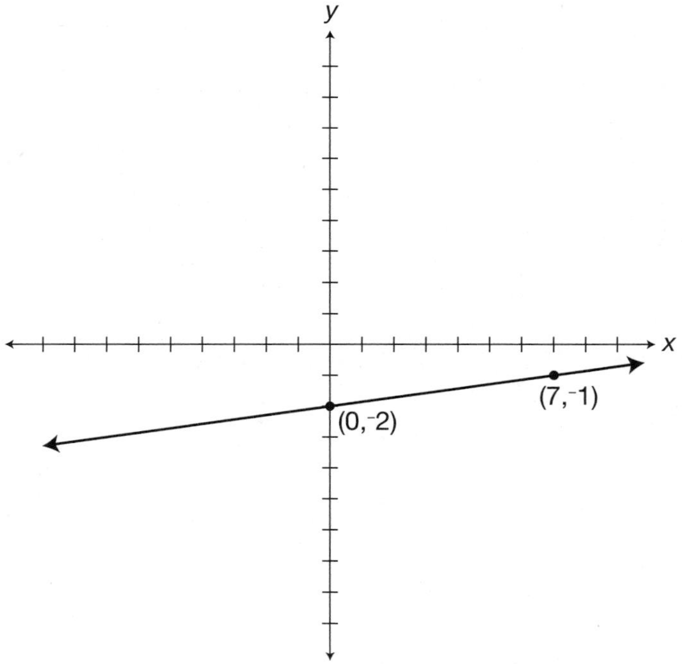

188. Exchange the terms on each side of the equal sign. $17y = {}^-34x + 85$

Divide both sides of the equation by 17. $\frac{17y}{17} = \frac{^-34x}{17} + \frac{85}{17}$

Simplify the equation. $y = {}^-2x + 5$

The equation is now in the proper slope/y-intercept form.

<u>slope:</u> $m = \frac{\text{change in } y}{\text{change in } x} = {}^-2 = \frac{^-2}{1}$

<u>y-intercept:</u> $b = 5$. The y-intercept is the point (0,5).

<u>Point on the line:</u> Starting at the point (0,5), a change in y of $^-2$ and in x of 1 gives the point $(0 + 1, 5 - 2)$ or (1,3).

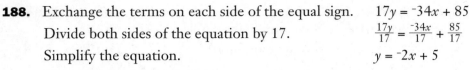

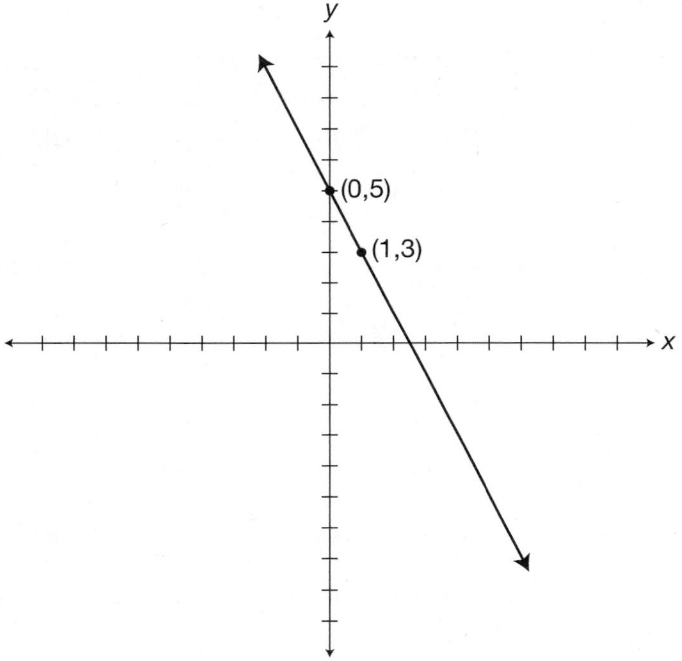

189. Put the equation in the proper slope/y-intercept form.

Add $3x$ to both sides of the equation.	$2y - 3x + 3x = 6 + 3x + 3x$
Simplify the equation.	$2y = 6 + 6x$
Divide both sides of the equation by 2.	$\frac{2y}{2} = \frac{6+6x}{2}$
Simplify the equation.	$y = \frac{6}{2} + \frac{6x}{2}$
	$y = 3 + 3x$
The equation is now in the proper slope/ y-intercept form.	$y = 3x + 3$

slope: $m = \frac{\text{change in } y}{\text{change in } x} = 3 = \frac{3}{1}$

y-intercept: $b = 3$, so the y-intercept is the point (0,3).

Point on the line: Starting at (0,3), a change in y of 3 and in x of 1 gives the point (0+1, 3+3) = (1, 6).

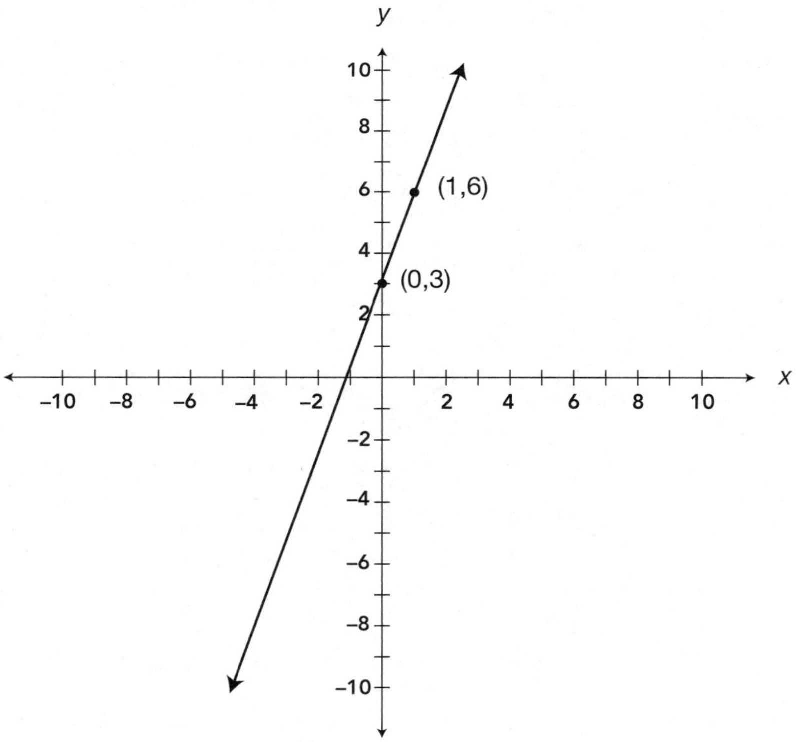

190. Put the equation in the proper slope/*y*-intercept form.

Use the distributive property of multiplication on the left side of the equation. Then, simplify.	$1 - (^{-}2x) - (3) = y + 3x$ $1 + 2x - 3 = y + 3x$ $2x - 2 = y + 3x$
Subtract $3x$ from both sides of the equation.	$2x - 3x - 2 = y + 3x - 3x$
Simplify the equation.	$^{-}x - 2 = y$
The equation is now in the proper slope/*y*-intercept form.	$y = -x - 2$

<u>slope:</u> $m = \frac{\text{change in } y}{\text{change in } x} = {}^{-}1 = \frac{^{-}1}{1}$

<u>*y*-intercept:</u> $b = {}^{-}2$, so the *y*-intercept is the point $(0,^{-}2)$.

<u>Point on the line:</u> Starting at $(0,^{-}2)$, a change in y of $^{-}1$ and in x of 1 gives the point $(0+1,^{-}2-1) = (1,^{-}3)$.

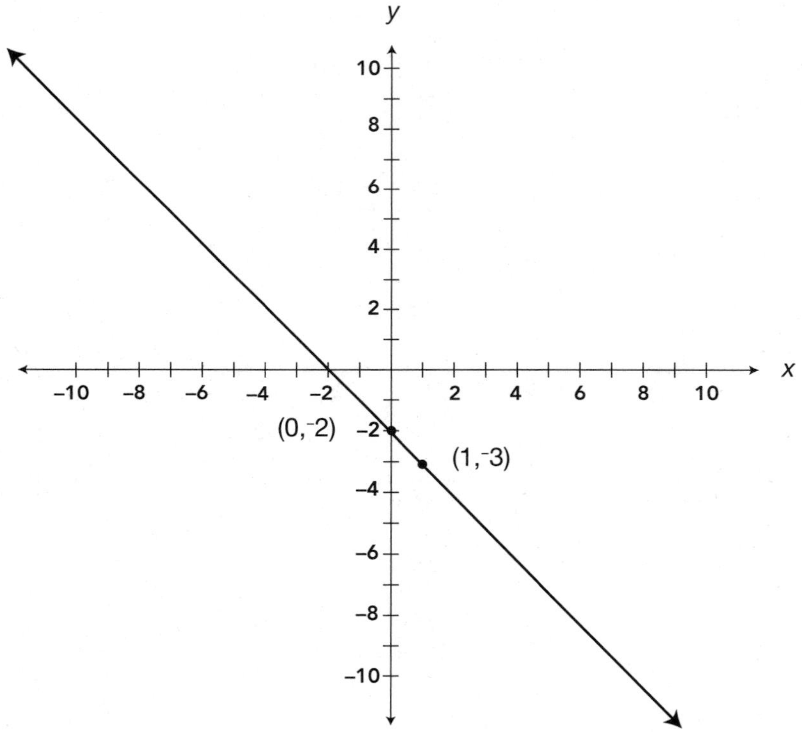

191. Put the equation in the proper slope/y-intercept form.

Use the distributive property of multiplication on multiplication on both sides of the equation. Then, simplify. $^-3x - y = 2y - 2(2x)$
$^-3x - y = 2y - 4x$

Add $4x$ to both sides of the equation. $^-3x + 4x - y = 2y - 4x + 4x$

Simplify the equation. $x - y = 2y$

Add y to both sides of the equation. $x - y + y = 2y + y$

Simplify the equation. $x = 3y$

Divide both sides of the equation by 3. $\frac{x}{3} = \frac{3y}{3}$

Simplify the equation. $\frac{x}{3} = y$

The equation is now in the proper slope/y-intercept form. $y = \frac{1}{3}x$

<u>slope:</u> $m = \frac{\text{change in } y}{\text{change in } x} = \frac{1}{3}$

<u>y-intercept:</u> $b = 0$, so the y-intercept is the point (0,0).

<u>Point on the line:</u> Starting at (0,0), a change in y of 1 and in x of 3 gives the point (0+3, 0+1) = (3,1).

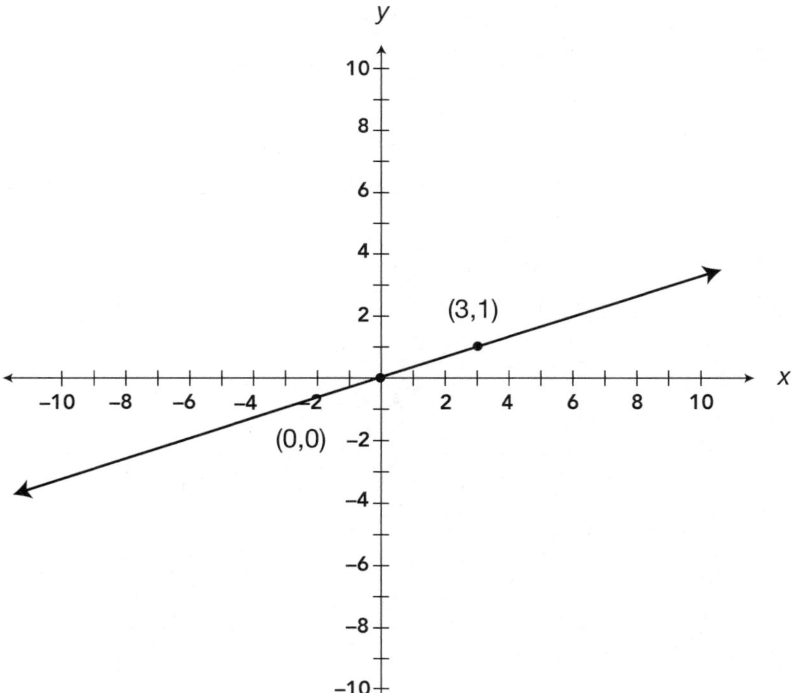

192. Put the equation in the proper slope/y-intercept form.

Use the distributive property of multiplication on the right-side of the equation. Then, simplify. $1 - x - 3y = {}^-2(3y) - 2(2x) - 2({}^-1)$

$1 - x - 3y = {}^-6y - 4x + 2$

Add x to both sides of the equation. $1 - x + x - 3y = {}^-6y - 4x + x + 2$

Simplify the equation. $1 - 3y = {}^-6y - 3x + 2$

Subtract 1 from both sides of the equation. $1 - 1 - 3y = {}^-6y - 3x + 2 - 1$

Simplify the equation. ${}^-3y = {}^-6y - 3x + 1$

Add 6 to both sides of the equation. $6y - 3y = 6y - 6y - 3x + 1$

Simplify the equation. $3y = {}^-3x + 1$

Divide both sides of the equation by 3. $\frac{3y}{3} = \frac{-3x+1}{3}$

Simplify the equation. $y = \frac{-3x}{3} + \frac{1}{3}$

The equation is now in the proper slope/y-intercept form. $y = {}^-x + \frac{1}{3}$

slope: $m = \frac{\text{change in } y}{\text{change in } x} = {}^-1 = \frac{{}^-1}{1}$

y-intercept: $b = \frac{1}{3}$, so the y-intercept is the point $(0, \frac{1}{3})$.

Point on the line: Starting at $(0, \frac{1}{3})$, a change in y of ${}^-1$ and in x of 1 gives the point $(0+1, {}^-1) = (1, {}^-\frac{2}{3})$.

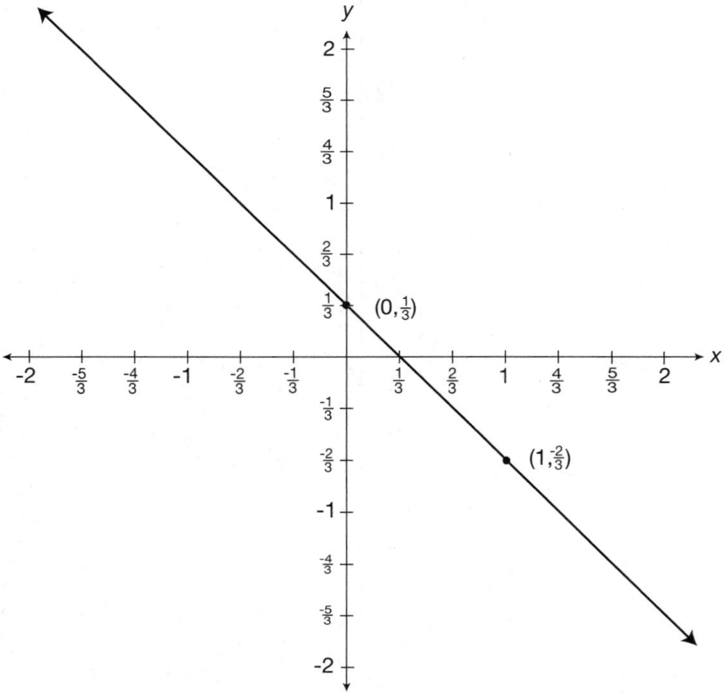

193. Put the equation in the proper slope/y-intercept form.

Use the distributive property of multiplication on the left side of the equation. Then, simplify.

$$\tfrac{2}{3}x - \tfrac{1}{9}y - (2) - \left(^{-}\tfrac{4}{3}x\right) = \tfrac{2}{9}y + 1$$
$$\tfrac{2}{3}x - \tfrac{1}{9}y - 2 + \tfrac{4}{3}x = \tfrac{2}{9}y + 1$$
$$\tfrac{2}{3}x + \tfrac{4}{3}x - \tfrac{1}{9}y - 2 = \tfrac{2}{9}y + 1$$
$$2x - \tfrac{1}{9}y - 2 = \tfrac{2}{9}y + 1$$

Subtract 1 from both sides of the equation. $2x - \tfrac{1}{9}y - 2 - 1 = \tfrac{2}{9}y + 1 - 1$

Simplify the equation. $2x - \tfrac{1}{9}y - 3 = \tfrac{2}{9}y$

Add $\tfrac{1}{9}y$ to both sides of the equation. $2x - \tfrac{1}{9}y + \tfrac{1}{9}y - 3 = \tfrac{2}{9}y + \tfrac{1}{9}y$

Simplify the equation. $2x - 3 = \tfrac{3}{9}y$

$2x - 3 = \tfrac{1}{3}y$

Multiply both sides of the equation by 3. $3(2x - 3) = 3\left(\tfrac{1}{3}y\right)$

Simplify the equation. $3(2x) + 3(^{-}3) = y$

$6x - 9 = y$

The equation is now in the proper slope/y-intercept form. $y = 6x - 9$

<u>slope:</u> $m = \dfrac{\text{change in } y}{\text{change in } x} = 6 = \dfrac{6}{1}$

<u>y-intercept:</u> $b = {}^{-}9$, so the y-intercept is the point $(0,{}^{-}9)$.

<u>Point on the line:</u> Starting at (0,9), a change in y of 6 and in x of 1 gives the point $(0+1,{}^{-}9 + 6) = (1,{}^{-}3)$.

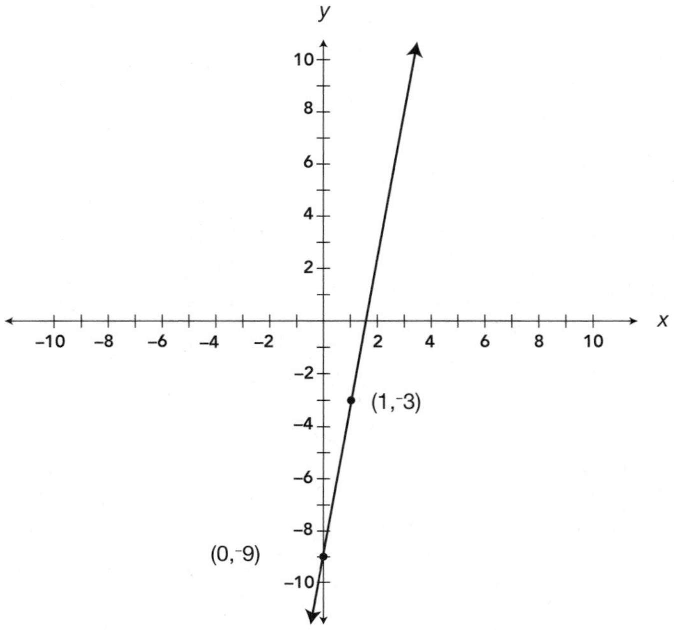

194. Put the equation in the proper slope/y-intercept form.

Multiply both sides of the equation by 5 (to clear the fractions).	$5 \times \frac{1}{5}({}^-y + 10x) = 5(y - x + 2)$ ${}^-y + 10x = 5(y - x + 2)$
Use the distributive property of multiplication on the right-side of the equation. Then, simplify.	${}^-y + 10x = 5(y) + 5({}^-x) + 5(2)$ ${}^-y + 10x = 5y - 5x + 10$
Add $5x$ to both sides of the equation.	${}^-y + 10x + 5x = 5y - 5x + 5x + 10$
Simplify the equation.	${}^-y + 15x = 5y + 10$
Subtract 10 from both sides of the equation.	${}^-y + 15x - 10 = 5y + 10 - 10$
Simplify the equation.	${}^-y + 15x - 10 = 5y$
Add y to both sides of the equation.	${}^-y + y + 15x - 10 = 5y + y$
Simplify the equation.	$15x - 10 = 6y$
Divide both sides of the equation by 6.	$\frac{15x-10}{6} = \frac{6y}{6}$ $\frac{15x}{6} - \frac{10}{6} = y$
Simplify the equation.	$\frac{5x}{2} - \frac{5}{3} = y$
The equation is now in the proper slope/y-intercept form.	$y = \frac{5}{2}x - \frac{5}{3}$

<u>slope:</u> $m = \frac{\text{change in } y}{\text{change in } x} = \frac{5}{2}$

<u>y-intercept:</u> $b = {}^-\frac{5}{3}$, so the y-intercept is the point $(0, {}^-\frac{5}{3})$.

<u>Point on the line:</u> Starting at $(0, {}^-\frac{5}{3})$, a change in y of 5 and in x of 2 gives the point $(0 + 2, {}^-\frac{5}{3} + 5) = (2, \frac{10}{3})$.

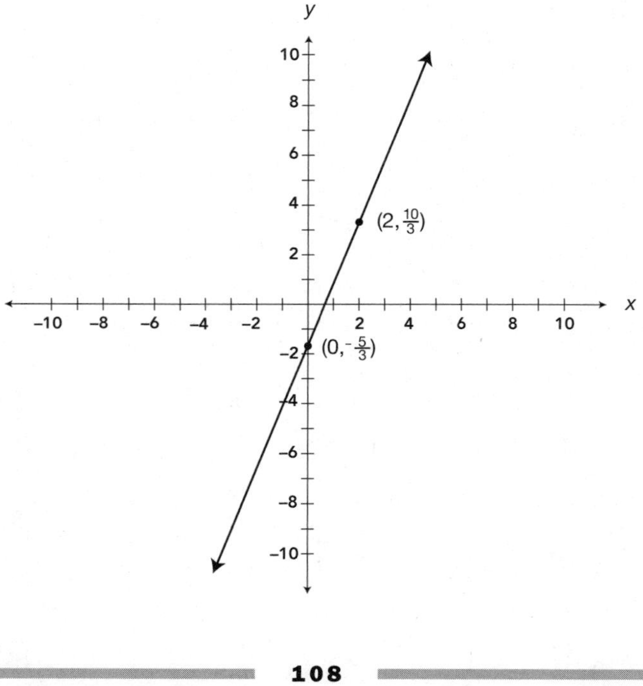

195. Put the equation in the proper slope/y-intercept form.

Remove the parentheses.	$0.1 - 0.5y - 0.4x + 0.4y + 0.3x + 0.4 = 0$
Use the commutative property of addition to gather like terms.	$0.1 + 0.4 - 0.5y + 0.4y - 0.4x + 0.3x = 0$
Simplify the equation.	$0.5 - 0.1y - 0.1x = 0$
Multiply both sides of the equation by 10 (to clear the decimals).	$5 - y - x = 0$
Add y to both sides of the equation.	$5 - y + y - x = 0 + y$
Simplify the equation.	$5 - x = y$
The equation is now in the proper slope/y-intercept form.	$y = {}^-x + 5$

<u>slope:</u> $m = \frac{\text{change in } y}{\text{change in } x} = {}^-1 = \frac{{}^-1}{1}$

<u>y-intercept:</u> $b = 5$, so the y-intercept is the point (0,5).

<u>Point on the line:</u> Starting at (0,5), a change in y of -1 and in x of 1 gives the point (0+1, 5–1) = (1,4).

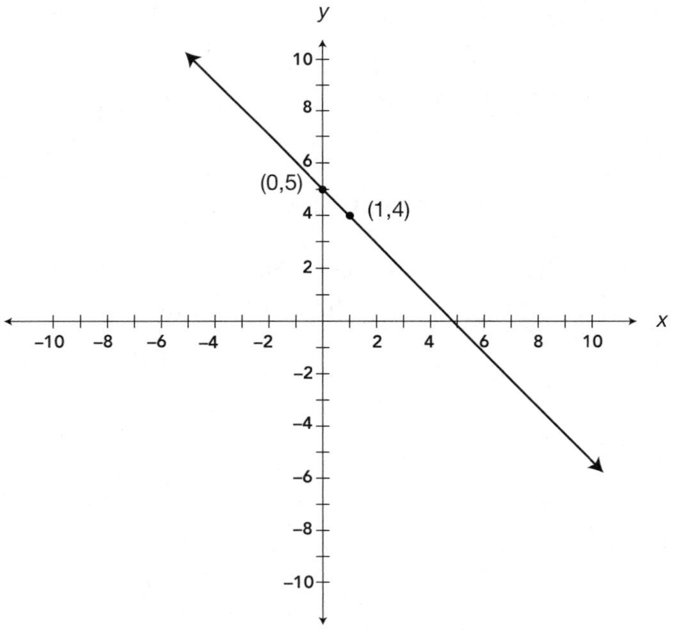

196. Let x = horizontal movement. Forward is in the positive direction.
Let y = vertical movement. Ascending is in the positive direction. Descending is in the negative.
The change in position of the glider is described by the slope.
The change in y is $^{-}1$ for every change in x of $^{+}25$.

slope: $m = \frac{\text{change in } y}{\text{change in } x} = \frac{-1}{25}$

The starting position for the purposes of this graphic solution is at an altitude of 250 ft or $^{+}250$. So: $b = 250$

Using the standard form $y = mx + b$, substitute the given values into the formula. $y = \frac{-1}{25}x + 250$

A graph of this equation would have a slope of $-\frac{1}{25}$ and the y-intercept would be (0,250).

197. Let y = the amount of a monthly bill.
Let x = the hours of Internet use for the month.
The costs for the month will equal \$15 plus the \$0.25 times the number of hours of use.
Written as an equation, this information would be as follows: $y = 0.25x + 15$

A graph of this equation would have a slope of 0.25 or $\frac{25}{100} = \frac{1}{4}$ the y-intercept would be (0,15).

198. Let y = the cost of a scooter rental for one day.
Let x = the number of miles driven in one day.
The problem tells us that the cost would be equal to the daily charges plus the 0.05 times the number of miles driven.
Written as an equation, this would be $y = 0.05x + 20$
The graph would have a y-intercept at (0,20) and the slope would be $\frac{5}{100} = \frac{1}{20}$.

199. Let y = the total cost for equipment.
Let x = the number of tanks used during the week.
The problem tells us that the cost would be equal to the weekly charge for gear rental plus 8 times the number of tanks used.
An equation that would represent this information would be $\underline{y = 8x + 150}$
The y-intercept would be (0,150) and the slope = $8 = \frac{8}{1}$.

200. Let y = the number of birds that visited a backyard feeder.
Let x = the number of chickadees that visited the feeder.
An equation that represents the statement would be: $\underline{y = 7x}$
The y-intercept is (0,0) and the slope = $7 = \frac{7}{1}$.

9

Solving Inequalities

If you compare Chapter 9 to Chapter 6, you will find only a few differences between solving inequalities and solving equations. The methods and procedures you use are virtually the same. Your goal is to isolate the variable on one side of the inequality, and the result will be your solution. In this chapter, there will be 25 inequalities so you can practice your solving skills. Look at the following tips to see what makes solving inequalities different from solving equalities.

Tips for Solving Inequalities

- Keep the inequality symbol facing the same way when you perform an addition or subtraction to both sides of the inequality. Keep the symbol facing the same way when you multiply or divide by a positive factor. But remember to **change** the direction of the inequality when you multiply or divide by a negative factor.
- When you have fractional coefficients or terms in an equation or inequality, multiply both sides by the least common multiple of the denominators and work with whole numbers. If the coefficients and/or terms are decimals, multiply by multiples of ten until you have whole numbers to work with.

Solve the following inequalities.

201. $3x + 2 < 11$

202. $4x - 6 > 30$

203. $\frac{2}{5}x \leq 18$

204. $4x + 26 \geq 90$

205. $8 - 6x < 50$

206. $^-2x + 18 \leq -26$

207. $2x + 0.29 > 0.79$

208. $^-6(x + 1) \geq 60$

209. $3(5 - 4x) < x - 63$

210. $4(x + 1) < 5(x + 2)$

211. $2(7x - 3) \geq {^-2}(5 + 3x)$

212. $^-3(2 - 4x) < 2\left(^-\frac{1}{2}x + 10\right)$

213. $\frac{-x}{0.3} \leq 20$

214. $\frac{4}{3}x - 5 > x - 2$

215. $2 - 3(1 - 3x) \geq 4 + 4\left(1 + \frac{5}{2}x\right)$

216. $^-4x + 3(x + 5) \geq 3(x + 2)$

217. $x - \frac{3}{4} < {^-\frac{3}{4}}(x + 2)$

218. $\frac{3}{2}x + 0.1 \geq 0.9 + x$

219. $1.2 - 0.2(1 - 3x) < 2.5 + 2(1 - x)$

220. $^-7(x + 3) < {^-4x}$

221. $\frac{5}{4}(x + 4) > \frac{1}{2}(x + 8) - 8$

222. $3(1 - 3x) \geq {^-3}(x + 27)$

223. $^-5[9 + (x - 4)] \geq 2(13 - x)$

224. $11(1 - x) \geq 3(3 - x) - 1$

225. $3(x - 16) - 2 < 9(x - 2) - 7x$

Answers

Numerical expressions in parentheses like this [ ] are operations performed on only part of the original expression. The operations performed within these symbols are intended to show how to evaluate the various terms that make up the entire expression.

Expressions with parentheses that look like this ( ) contain either numerical substitutions or expressions that are part of a numerical expression. Once a single number appears within these parentheses, the parentheses are no longer needed and need not be used the next time the entire expression is written.

When two pair of parentheses appear side by side like this ( )( ), it means that the expressions within are to be multiplied.

Sometimes parentheses appear within other parentheses in numerical or algebraic expressions. Regardless of what symbol is used, ( ), { }, or [ ], perform operations in the innermost parentheses first and work outward.

Underlined inequalities show the simplified result.

201.	Subtract 2 from both sides of the inequality.	$3x + 2 - 2 < 11 - 2$
	Simplify the inequality.	$3x < 9$
	Divide both sides of the inequality by 3.	$\frac{3x}{3} < \frac{9}{3}$
	Simplify.	$\underline{x < 3}$
202.	Add 6 to both sides of the inequality.	$4x - 6 + 6 > 30 + 6$
	Simplify the inequality.	$4x > 36$
	Divide both sides of the inequality by 4.	$\frac{4x}{4} > \frac{36}{4}$
	Simplify.	$\underline{x > 9}$
203.	Multiply both sides of the inequality by $\frac{5}{2}$.	$\frac{5}{2}(\frac{2}{5})x \leq \frac{5}{2}(18)$
	Simplify.	$(\frac{10}{10})x \leq \frac{5}{2}(\frac{18}{1})$
		$(1)x \leq 45$
		$\underline{x \leq 45}$
204.	Subtract 26 from both sides of the inequality.	$4x + 26 - 26 \geq 90 - 26$
	Simplify.	$4x \geq 64$
	Divide both sides of the inequality by 4.	$\frac{4x}{4} \geq \frac{64}{4}$
	Simplify the inequality.	$\underline{x \geq 16}$
205.	Subtract 8 from both sides of the inequality.	$8 - 8 - 6x < 50 - 8$
	Simplify the inequality.	$^-6x < 42$
	Divide both sides of the inequality by $^-6$ and change the direction of the inequality sign.	$\frac{^-6x}{^-6} > \frac{42}{^-6}$
	Simplify.	$\underline{x > {}^-7}$

206. Subtract 18 from both sides of the inequality. $^-2x + 18 - 18 \leq {}^-26 - 18$
Combine like terms on each side of the inequality. $^-2x \leq {}^-44$
Divide both sides of the inequality by -2. (Make certain to reverse the inequality sign because you are dividing by a negative number.) $\frac{^-2x}{^-2} \geq \frac{^-44}{^-2}$
Simplify the inequality. $\underline{x \geq 22}$

207. Subtract 0.29 from both sides of the inequality. $2x + 0.29 - 0.29 > 0.79 - 0.29$
Combine like terms on each side of the inequality. $2x > 0.50$
Divide both sides of the inequality by 2. $\frac{2x}{2} > \frac{0.50}{2}$
Simplify the inequality. $\underline{x > 0.25}$

208. Divide both sides of the inequality by $^-6$ and change the direction of the inequality sign. $\frac{^-6(x+1)}{^-6} \leq \frac{60}{^-6}$
Simplify the expressions. $x + 1 \leq {}^-10$
Subtract 1 from both sides of the inequality. $x + 1 - 1 \leq {}^-10 - 1$
Simplify. $\underline{x \leq {}^-11}$

209. Use the distributive property of multiplication. $3(5) - 3(4x) < x - 63$
Simplify. $15 - 12x < x - 63$
Add $12x$ to both sides of the inequality. $15 - 12x + 12x < x + 12x - 63$
Combine like terms on each side of the inequality. $15 < 13x - 63$
Add 63 to both sides of the inequality. $15 + 63 < 13x - 63 + 63$
Simplify the inequality. $78 < 13x$
Divide both sides of the inequality by 13. $\frac{78}{13} < \frac{13x}{13}$
Simplify. $6 < x$
If $a < b$, then $b > a$. $\underline{x > 6}$

210. Use the distributive property of multiplication. $4(x) + 4(1) < 5(x) + 5(2)$
Simplify. $4x + 4 < 5x + 10$
Subtract 4 from both sides of the inequality. $4x + 4 - 4 < 5x + 10 - 4$
Combine like terms on each side of the inequality. $4x < 5x + 6$
Subtract $5x$ from both sides of the inequality. $4x - 5x < 5x - 5x + 6$
Simplify the inequality. $^-x < 6$
Multiply both sides of the equation by $^-1$ and change the direction of the inequality sign. $^-1(^-x) > {}^-1(6)$
Simplify. $\underline{x > {}^-6}$

211. Use the distributive property of multiplication. $2(7x) - 2(3) \geq {}^-2(5) - 2(3x)$
Simplify the expressions. $14x - 6 \geq {}^-10 - 6x$
Add $6x$ to both sides of the inequality. $14x + 6x - 6 \geq {}^-10 - 6x + 6x$
Combine like terms. $20x - 6 \geq -10$
Add 6 to both sides of the inequality. $20x - 6 + 6 \geq {}^-10 + 6$
Simplify. $20x \geq {}^-4$
Divide both sides of the inequality by 20. $\frac{20x}{20} \geq \frac{^-4}{20}$
Simplify. $x \geq \frac{^-4}{20}$
Reduce the fraction to lowest terms. $\underline{x \geq \frac{^-1}{5}}$

212. Use the distributive property of multiplication on both sides of the inequality. $^-3(2) - 3(^-4x) < 2\left(^-\frac{1}{2}x\right) + 2(10)$
$^-6 + 12x < {}^-x + 20$
Add 6 to both sides of the inequality. $^-6 + 6 + 12x < {}^-x + 20 + 6$
Combine like terms on each side of the inequality. $12x < {}^-x + 26$
Add x to both sides of the inequality. $12x + x < {}^-x + x + 26$
Combine like terms on each side of the inequality. $13x < 26$
Divide both sides of the inequality by 13. $\frac{13x}{13} < \frac{26}{13}$
Simplify the inequality. $\underline{x < 2}$

213. Multiply both sides of the inequality by 0.3. $0.3(\frac{^-x}{0.3}) \leq 0.3(20)$

Simplify the expressions on both sides. $^-x \leq 6$

Multiply both sides of the inequality by $^-1$ and change the direction of the inequality sign. $^-1(^-x) \geq {^-1}(6)$

Simplify the expressions. $\underline{x \geq {^-6}}$

214. Add 5 to both sides of the inequality. $\frac{4}{3}x - 5 + 5 > x - 2 + 5$

Simplify. $\frac{4}{3}x > x + 3$

Subtract $1x$ from both sides of the inequality. $\frac{4}{3}x - x > x - x + 3$

$\frac{4}{3}x - \frac{3}{3}x > x - x + 3$

Simplify the expressions. $\frac{1}{3}x > 3$

Multiply both sides of the inequality by 3. $3(\frac{1}{3}x) > 3(3)$

$\frac{3}{1}(\frac{1}{3}x) > 3\ (3)$

Simplify the expressions. $\underline{x > 9}$

215. Use the distributive property of multiplication on both sides of the inequality. $2 - 3(1) - 3(^-3x) \geq 4 + 4(1) + 4(\frac{5}{2}x)$

$2 - 3 + 9x \geq 4 + 4 + 10x$

$^-1 + 9x \geq 8 + 10x$

Subtract 8 from both sides of the inequality. $^-1 - 8 + 9x \geq 8 - 8 + 10x$

Combine like terms on each side of the inequality. $^-9 + 9x \geq 10x$

Subtract $9x$ from both sides of the inequality. $^-9 + 9x - 9x \geq 10x - 9x$

Combine like terms on each side of the inequality. $^-9 \geq x$

If $a \geq b$, then $b \leq a$. $\underline{x \leq {^-9}}$

216. Use the distributive property of multiplication. $^{-}4x + 3(x) + 3(5) \geq 3(x) + 3(2)$

Simplify. $^{-}4x + 3x + 15 \geq 3x + 6$

Combine like terms. $(^{-}4x + 3x) + 15 \geq 3x + 6$

Simplify. $^{-}x + 15 \geq 3x + 6$

Add x to both sides of the inequality. $x - x + 15 \geq x + 3x + 6$

Combine like terms. $15 \geq 4x + 6$

Subtract 6 from both sides of the inequality. $15 - 6 \geq 4x + 6 - 6$

Simplify. $9 \geq 4x$

Divide both sides of the inequality by 4. $\frac{9}{4} \geq \frac{4x}{4}$

Simplify. $\frac{9}{4} \geq x$

Express the fraction in its simplest form. $2\frac{1}{4} \geq x$

If $a \geq b$, then $b \leq a$. $\underline{x \leq 2\frac{1}{4}}$

217. You can simplify inequalities with fractions by multiplying them by a common multiple of the denominators.

Multiply both sides of the inequality by 4. $4(x - \frac{3}{4}) < 4(\frac{^{-}3}{4}(x + 2))$

Use the distributive property of multiplication. $4(x) - 4(\frac{3}{4}) < 4(\frac{^{-}3}{4})(x + 2)$

Simplify the expressions. $4x - 3 < {}^{-}3(x + 2)$

Use the distributive property of multiplication. $4x - 3 < {}^{-}3(x) - 3(2)$

Simplify. $4x - 3 < {}^{-}3x - 6$

Add 3 to both sides of the inequality. $4x - 3 + 3 < {}^{-}3x - 6 + 3$

Combine like terms. $4x < -\ 3x - 3$

Add $3x$ to both sides of the equation. $3x + 4x < 3x - 3x - 3$

Combine like terms. $7x < {}^{-}3$

Divide both sides of the inequality by 7. $\frac{7x}{7} < \frac{^{-}3}{7}$

Simplify the expressions. $\underline{x < \frac{^{-}3}{7}}$

218. Subtract 0.1 from both sides of the inequality. $\frac{3}{2}x + 0.1 - 0.1 \geq 0.9 - 0.1 + x$

Combine like terms on each side of the inequality. $\frac{3}{2}x \geq 0.8 + x$

Subtract x from both sides of the inequality. $\frac{3}{2}x - x \geq 0.8 + x - x$

Simplify. $\frac{1}{2}x \geq 0.8$

Multiply both sides of the inequality by 2. $2(\frac{1}{2}x) \geq 2(0.8)$

Simplify the expressions. $\underline{x \geq 1.6}$

219. Use the distributive property of multiplication on both sides of the inequality. $1.2 - 0.2(1) - 0.2(-3x) < 2.5 + 2(1) + 2(-x)$

$1.2 - 0.2 + 0.6x < 2.5 + 2 - 2x$

$1 + 0.6x < 4.5 - 2x$

Subtract 1 from both sides of the inequality. $1 - 1 + 0.6x < 4.5 - 1 - 2x$

Combine like terms on each side of the inequality. $0.6x < 3.5 - 2x$

Add $2x$ to both sides of the inequality. $0.6x + 2x < 3.5 - 2x + 2x$

Combine like terms on each side of the inequality. $2.6x < 3.5$

Divide both sides of the inequality by 2.6. $\frac{2.6x}{2.6} < \frac{3.5}{2.6}$

$x < \frac{3.5}{2.6}$

Simplify the inequality. $x < \frac{35}{26} = 1\frac{9}{26}$

$\underline{x < 1\frac{9}{26}}$

220. Use the distributive property of multiplication. $^-7(x) - 7(3) < {}^-4x$

Simplify the terms. $^-7x - 21 < {}^-4x$

Add 21 to both sides of the inequality. $^-7x - 21 + 21 < {}^-4x + 21$

Simplify by combining like terms. $^-7x < {}^-4x + 21$

Add $4x$ to both sides of the inequality. $^-7x + 4x < {}^-4x + 4x + 21$

Simplify the terms. $^-3x < 21$

Divide both sides of the inequality by $^-3$ and change the direction of the inequality sign. $\frac{^-3x}{^-3} > \frac{21}{^-3}$

Simplify the expressions. $\underline{x > {}^-7}$

221. Use the distributive property of multiplication. $\frac{5}{4}(x) + \frac{5}{4}(4) > \frac{1}{2}(x) + \frac{1}{2}(8) - 8$

Simplify the terms. $\frac{5}{4}x + 5 > \frac{1}{2}x + 4 - 8$

Combine like terms. $\frac{5}{4}x + 5 > \frac{1}{2}x - 4$

Subtract $\frac{1}{2}x$ from both sides of the inequality. $\frac{5}{4}x - \frac{1}{2}x + 5 > \frac{1}{2}x - \frac{1}{2}x - 4$

Combine like terms. $\frac{3}{4}x + 5 > {}^-4$

Subtract 5 from both sides of the inequality. $\frac{3}{4}x + 5 - 5 > {}^-4 - 5$

Simplify. $\frac{3}{4}x > {}^-9$

Multiply both sides by $\frac{4}{3}$ (the reciprocal of $\frac{3}{4}$). $\frac{4}{3}(\frac{3}{4}x) > \frac{4}{3}(^-9)$

Simplify the expressions. $\underline{x > {}^-12}$

222. Use the distributive property of multiplication. $3(1) - 3(3x) \geq {}^-3(x) - 3(27)$
Simplify terms. $3 - 9x \geq {}^-3x - 81$
Add $9x$ to both sides. $3 - 9x + 9x \geq 9x - 3x - 81$
Combine like terms. $3 \geq 6x - 81$
Add 81 to both sides of the inequality. $3 + 81 \geq 6x - 81 + 81$
Combine like terms. $84 \geq 6x$
Divide both sides of the inequality by 6. $\frac{84}{6} \geq \frac{6x}{6}$
Simplify. $14 \geq x$
If $a \geq b$, then $b \leq a$. $\underline{x \leq 14}$

223. Remove the inner brackets, use the commutative property of addition and combine like terms. $^-5[9 + x - 4] \geq 2(13 - x)$
$^-5[x + 9 - 4] \geq 2(13 - x)$
$^-5[x + 5] \geq 2(13 - x)$
Use the distributive property of multiplication. $^-5[x] - 5[5] \geq 2(13) - 2(x)$
Simplify terms. $^-5x - 25 \geq 26 - 2x$
Add $5x$ to both sides of the inequality. $^-5x + 5x - 25 \geq 26 - 2x + 5x$
Combine like terms. $^-25 \geq 26 + 3x$
Subtract 26 from both sides of the inequality. $^-25 - 26 \geq 26 - 26 + 3x$
Combine like terms. $^-51 \geq 3x$
Divide both sides of the inequality by 3. $\frac{^-51}{3} \geq \frac{3x}{3}$
Simplify the expressions. $^-17 \geq x$
If $a \geq b$, then $b \leq a$. $\underline{x \leq {}^-17}$

224. Use the distributive property of multiplication. $11(1) - 11(x) \geq 3(3) - 3(x) - 1$
Simplify terms. $11 - 11x \geq 9 - 3x - 1$
Use the commutative property. $11 - 11x \geq 9 - 1 - 3x$
Combine like terms. $11 - 11x \geq 8 - 3x$
Subtract 8 from both sides of the inequality. $11 - 8 - 11x \geq 8 - 8 - 3x$
Combine like terms. $3 - 11x \geq {}^-3x$
Add $11x$ to both sides. $3 - 11x + 11x \geq {}^-3x + 11x$
Combine like terms. $3 \geq 8x$
Divide both sides by 8. $\frac{3}{8} \geq \frac{8x}{8}$
Simplify the expressions. $\frac{3}{8} \geq x$
If $a \geq b$, then $b \leq a$. $\underline{x \leq \frac{3}{8}}$

225. Use the distributive property of multiplication.	$3(x) - 3(16) - 2 < 9(x) - 9(2) - 7x$
Simplify terms.	$3x - 48 - 2 < 9x - 18 - 7x$
Use the commutative property to associate like terms.	$3x - 48 - 2 < 9x - 7x - 18$
Simplify terms.	$3x - 50 < 2x - 18$
Add 50 to both sides of the inequality.	$3x - 50 + 50 < 2x - 18 + 50$
Combine like terms.	$3x < 2x + 32$
Subtract $2x$ from both sides of the inequality.	$3x - 2x < 2x - 2x + 32$
Combine like terms.	$\underline{x < 32}$

10

Graphing Inequalities

In this chapter, you will practice graphing the solutions to inequalities involving one or two variables. When there is only one variable, you use a number line. When there are two variables, you use a coordinate plane.

Tips for Graphing Solutions to Inequalities

When using a number line to illustrate the solution for an inequality, use a solid circle on the number line as the endpoint when the inequality symbol is $\leq$ or $\geq$. When the inequality symbol is $<$ or $>$, use an open circle to show the endpoint. A solid circle indicates that the solution graph includes the endpoint; an open circle indicates that the solution graph does not include the endpoint.

When there are two variables, use a coordinate plane to graph the solution. Use the skills you have been practicing in the previous chapters to transform the inequality into the slope/y-intercept form you used to graph equations with two variables. Use the slope and the y-intercept of the transformed inequality to show the boundary line for your solution graph. Draw a solid line when the inequality symbol is $\leq$ or $\geq$. Draw a dashed line when the inequality symbol is $<$ or $>$. To complete the graph, shade the region above the boundary line when the inequality symbol is $>$ or $\geq$. Shade the region below the boundary line when the inequality symbol is $<$ or $\leq$.

A simple way to check your graphic solution is to pick a point on either side of the boundary line and substitute the x and y values in your inequality. If the result is a true statement, then you have shaded the correct side of the boundary line. A convenient point to use, if it is not your y-intercept, is the origin (0,0).

Graph the solutions of the following inequalities on a number line.

226. $x \geq 4$

227. $x < {}^-1$

228. $x \leq 6$

229. $x < 5$

230. $x > {}^-1$

Graph the solutions of the following inequalities on a coordinate plane. (Use graph paper.)

231. $y < x + 1$

232. $y \geq {}^-x + 2$

233. $y > {}^-3x - 2$

234. $\frac{1}{2}x + y \leq 3$

235. $2y - 3x < 8$

236. $y + 2 \leq 3x + 5$

237. $3x - 4 \geq 2y$

238. $\frac{3}{4}y + 6 \geq 3x$

239. $0.5y - x + 3 > 0$

240. $1.4x - 0.7y \leq 2.8$

241. $\frac{y}{3} < \frac{2}{3} - x$

242. ${}^-\frac{2}{9}y + \frac{2}{3}x \leq {}^-2$

243. $0.5x > 0.3\,y - 0.9$

244. $3x - y \leq 7x + y - 8$

245. ${}^-(2 - 3y) < 1 + 3\,(2 - x)$

246. ${}^-12 \leq {}^-3(x + y)$

247. $9y + 7 \geq 2(x + 8)$

248. $2(y + 3) - x \geq 6(1 - x)$

249. ${}^-28y \geq 2x - 14(y + 10)$

250. ${}^-2[1 - 2(2y - 3x)] \leq 0$

Answers

Numerical expressions in parentheses like this [ ] are operations performed on only part of the original expression. The operations performed within these symbols are intended to show how to evaluate the various terms that make up the entire expression.

Expressions with parentheses that look like this ( ) contain either numerical substitutions or expressions that are part of a numerical expression. Once a single number appears within these parentheses, the parentheses are no longer needed and need not be used the next time the entire expression is written.

When two pair of parentheses appear side by side like this ( )( ), it means that the expressions within are to be multiplied.

Sometimes parentheses appear within other parentheses in numerical or algebraic expressions. Regardless of what symbol is used, ( ), { }, or [ ], perform operations in the innermost parentheses first and work outward.

The answers to these questions are the graphs.

226.

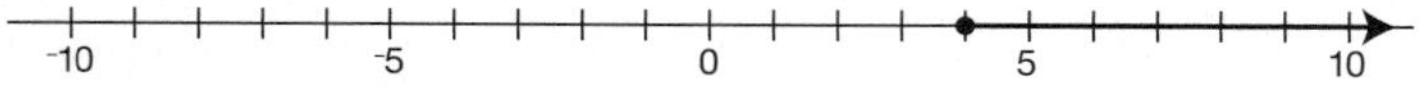

227.

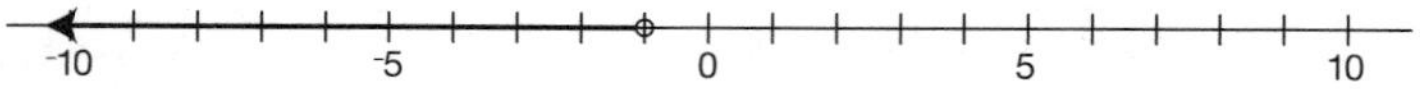

228.

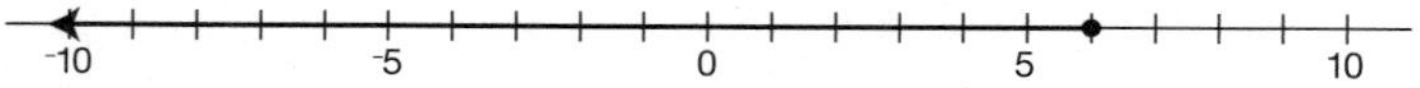

229.

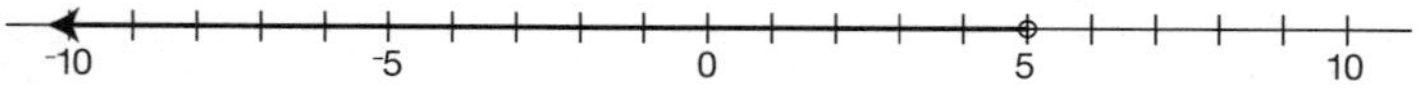

230.

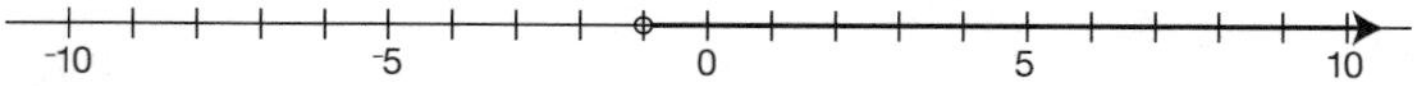

231. The inequality is now in the proper slope/y-intercept form.

slope: $m = \frac{\text{change in } y}{\text{change in } x} = 1 = \frac{1}{1}$

y-intercept: $b = 1$. The y-intercept is the point (0,1).

Point on the line: Starting at (0,1), a change in y of 1 and in x of 1 gives the point (0 + 1,1 + 1) or (1,2).

Draw a dotted boundary line and shade below it.

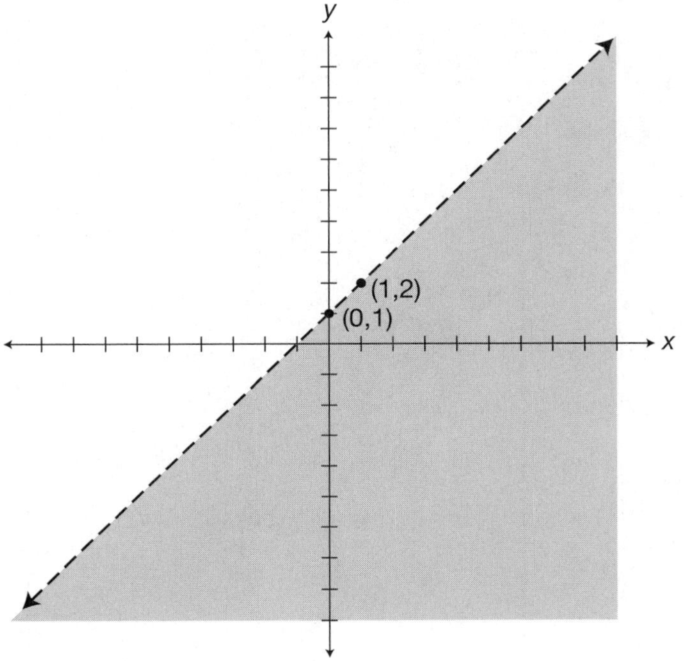

232. The inequality is now in the proper slope/*y*-intercept form.

<u>slope:</u> $m = \frac{\text{change in } y}{\text{change in } x} = {}^-1 = \frac{-1}{1}$

<u>*y*-intercept:</u> $b = 2$. The *y*-intercept is the point (0,2).

<u>Point on the line:</u> Starting at (0,2), a change in *y* of $^-1$ and in *x* of 1 gives the point (0+1,2−1) or (1,1).

Draw a solid boundary line and shade above it.

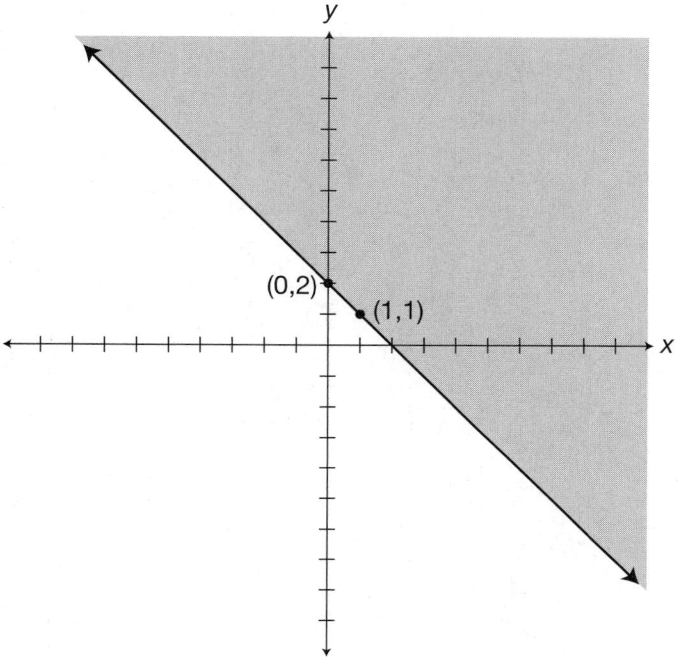

233. The inequality is now in the proper slope/*y*-intercept form.

slope: $m = \frac{\text{change in } y}{\text{change in } x} = {}^-3 = \frac{{}^-3}{1}$

y-intercept: $b = {}^-2$. The *y*-intercept is the point $(0,{}^-2)$.

Point on the line: Starting at $(0,{}^-2)$, a change in *y* of $^-3$ and in *x* of 1 gives the point $(0+1,{}^-2-3)$ or $(1,{}^-5)$.

Draw a dotted boundary line and shade below it.

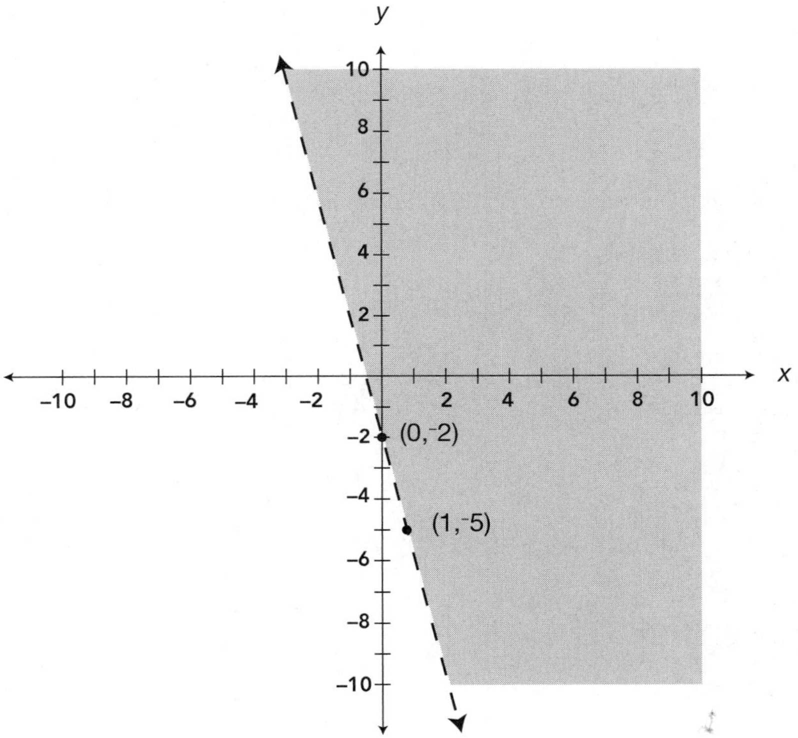

234. Subtract $\frac{1}{2}x$ from both sides of the inequality. $\frac{1}{2}x - \frac{1}{2}x + y \le {}^{-}\frac{1}{2}x + 3$
Combine like terms. $y \le {}^{-}\frac{1}{2}x + 3$
The inequality is now in the proper slope/y-intercept form.

slope: $m = \frac{\text{change in } y}{\text{change in } x} = \frac{^{-}1}{2}$

y-intercept: $b = 3$. The y-intercept is the point (0,3).

Point on the line: Starting at (0,3), a change in y of $^{-}1$ and in x of 2 gives the point (0+2,3−1) or (2,2).

Draw a solid boundary line and shade below it.

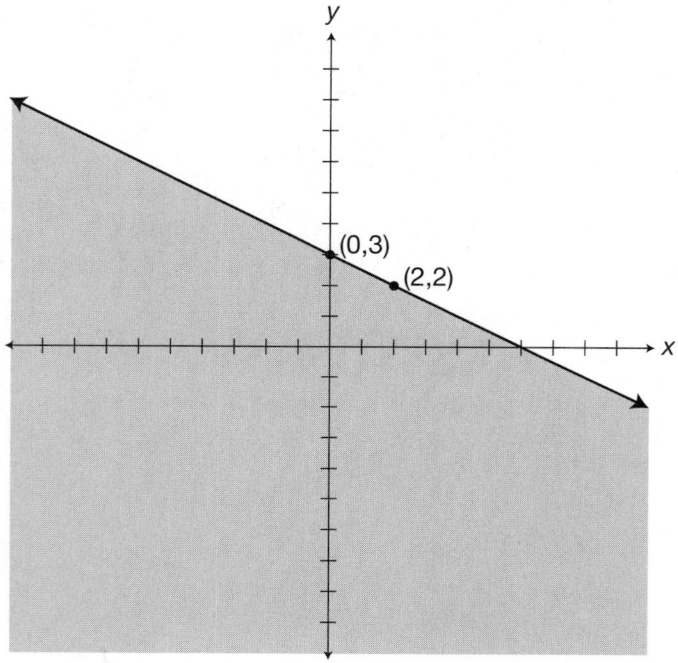

235. Add $3x$ to both sides of the inequality. $2y - 3x + 3x < 3x + 8$
Combine like terms. $2y < 3x + 8$

Divide both sides of the inequality by 2. $\frac{2y}{2} < \frac{3x}{2} + \frac{8}{2}$

Simplify terms. $y < \frac{3}{2}x + 4$

The inequality is now in the proper slope/y-intercept form.

slope: $m = \frac{\text{change in } y}{\text{change in } x} = \frac{3}{2}$

y-intercept: $b = 4$. The y-intercept is the point (0,4).

Point on the line: Starting at (0,4), a change in y of 3 and in x of 2 gives the point (0+2,4+3) or (2,7).

Draw a dotted boundary line and shade below it.

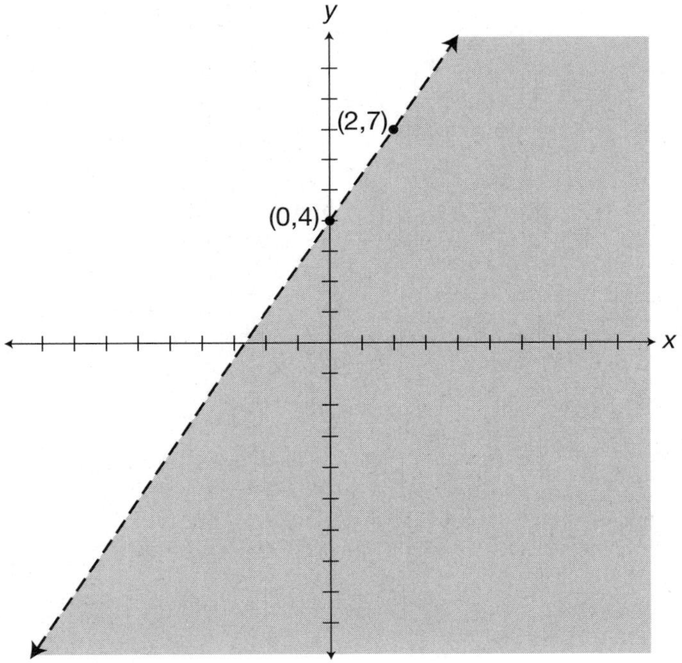

236. Subtract 2 from both sides of the inequality. $y + 2 - 2 \leq 3x + 5 - 2$
Combine like terms. $y \leq 3x + 3$
The inequality is now in the proper slope/y-intercept form.

slope: $m = \frac{\text{change in } y}{\text{change in } x} = 3 = \frac{3}{1}$

y-intercept: $b = 3$. The y-intercept is the point (0,3).

Point on the line: Starting at (0,3), a change in y of 3 and in x of 1 gives the point (0+1,3+3) or (1,6).

Draw a solid boundary line and shade below it.

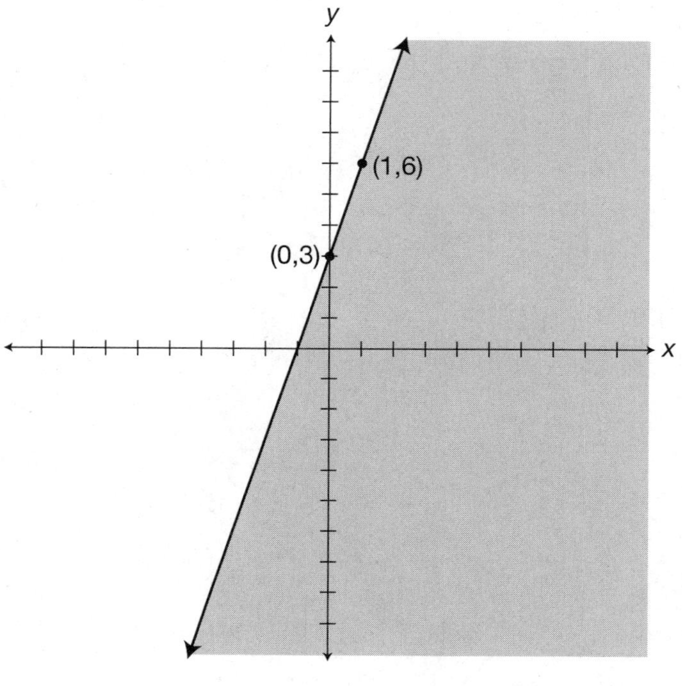

237. In an equation, if $c = d$, then $d = c$. But for an inequality, the direction of the inequality symbol must change when you change sides of the statement. If $c \geq d$, then $d \leq c$. Rewrite the inequality with sides exchanged and the symbol reversed. $2y \leq 3x - 4$

Divide both sides of the inequality by 2. $\frac{2y}{2} \leq \frac{3x}{2} - \frac{4}{2}$

Simplify terms. $y \leq \frac{3}{2}x - 2$

The inequality is now in the proper slope/y-intercept form.

slope: $m = \frac{\text{change in } y}{\text{change in } x} = \frac{3}{2}$

y-intercept: $b = {}^-2$. The y-intercept is the point $(0,{}^-2)$.

Point on the line: Starting at $(0,{}^-2)$, a change in y of 3 and in x of 2 gives the point $(0+2,{}^-2+3)$ or $(2,1)$.

Draw a solid boundary line and shade below it.

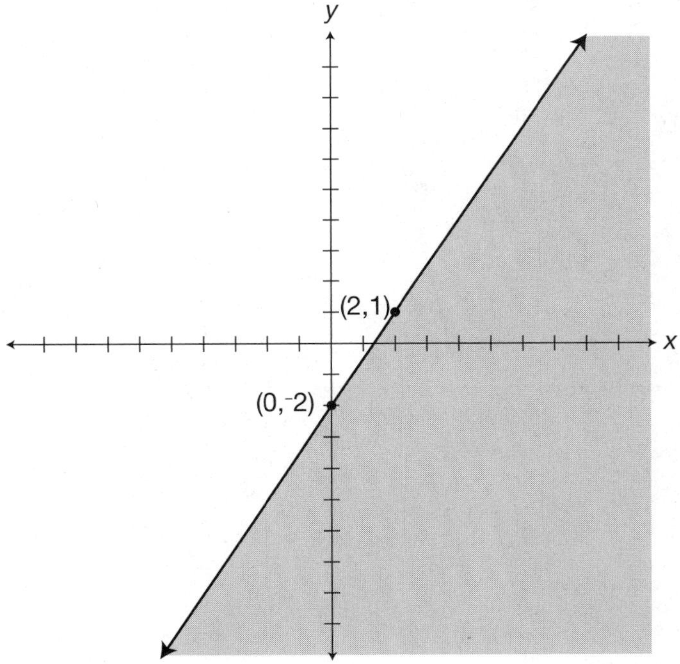

238. Subtract 6 from both sides of the inequality. $\frac{3}{4}y + 6 - 6 \geq 3x - 6$

Combine like terms. $\frac{3}{4}y \geq 3x - 6$

Multiply both sides of the inequality by the reciprocal $\frac{4}{3}$. $\frac{4}{3}(\frac{3}{4}y) \geq \frac{4}{3}(3x - 6)$

Use the distributive property of multiplication. $\frac{4}{3}(\frac{3}{4}y) \geq \frac{4}{3}(3x) - \frac{4}{3}(6)$

Simplify terms. $y \geq 4x - 8$

The inequality is now in the proper slope/y-intercept form.

<u>slope:</u> $m = \frac{\text{change in } y}{\text{change in } x} = \frac{4}{1}$

<u>y-intercept:</u> $b = {}^-8$. The y-intercept is the point $(0,{}^-8)$.

<u>Point on the line:</u> Starting at $(0,{}^-8)$, a change in y of 4 and in x of 1 gives the point $(0+1,{}^-8+4)$ or $(1,{}^-4)$.

Draw a solid boundary line and shade above it.

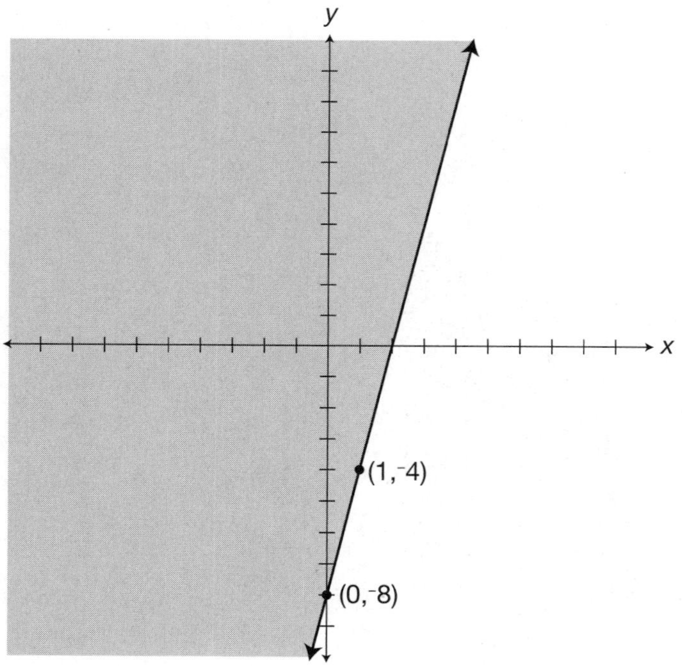

239. Subtract 3 from both sides of the inequality. $0.5y - x + 3 - 3 > 0 - 3$

Combine like terms on each side of the inequality. $0.5y - x > {}^{-}3$

Add $1x$ to both sides of the inequality. $0.5y + x - x > x - 3$

Combine like terms. $0.5y > x - 3$

Divide both sides of the inequality by 0.5. $\frac{0.5y}{0.5} > \frac{x}{0.5} - \frac{3}{0.5}$

Simplify the expressions. $y > \frac{x}{0.5} - \frac{3}{0.5}$

Simplify terms. $y > 2x - 6$

The inequality is now in the proper slope/y-intercept form.

<u>slope:</u> $m = \frac{\text{change in } y}{\text{change in } x} = 2 = \frac{2}{1}$

<u>y-intercept:</u> $b = {}^{-}6$. The y-intercept is the point $(0,{}^{-}6)$.

<u>Point on the line:</u> Starting at $(0,{}^{-}6)$, a change in y of 2 and in x of 1 gives the point $(0+1,{}^{-}6+2)$ or $(1,{}^{-}4)$.

Draw a dotted boundary line and shade above it.

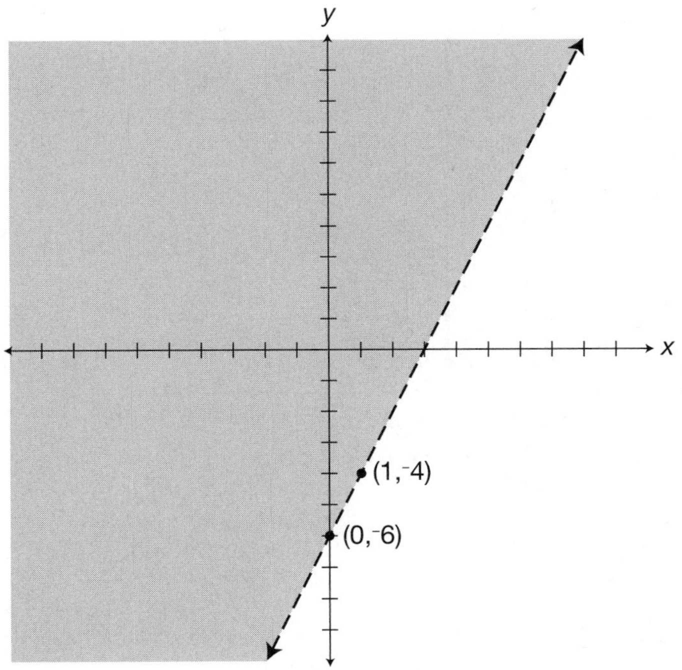

240. Put the inequality in the proper slope/y-intercept form.

Add $0.7y$ to both sides of the inequality.	$1.4x - 0.7y + 0.7y \leq 2.8 + 0.7y$
Combine like terms on the left side of the inequality.	$1.4x \leq 2.8 + 0.7y$
Subtract 2.8 from both sides of the inequality.	$1.4x - 2.8 \leq 2.8 - 2.8 + 0.7y$
Combine like terms on the right-side of the inequality.	$1.4x - 2.8 \leq 0.7y$
Divide both sides of the inequality by 0.7.	$\frac{1.4x-2.8}{0.7} \leq \frac{0.7y}{0.7}$
Simplify the inequality.	$2x - 4 \leq y$
If $b \leq a$, then $a \geq b$.	$y \geq 2x - 4$

<u>slope:</u> $m = \frac{\text{change in } y}{\text{change in } x} = 2 = \frac{2}{1}$

<u>y-intercept:</u> b = -4, so the y-intercept is the point (0,-4).

<u>Point on the line:</u> Starting at (0,-4), a change in y of 2 and in x of 1 gives the point (0+1,-4+2) = (1,-2).

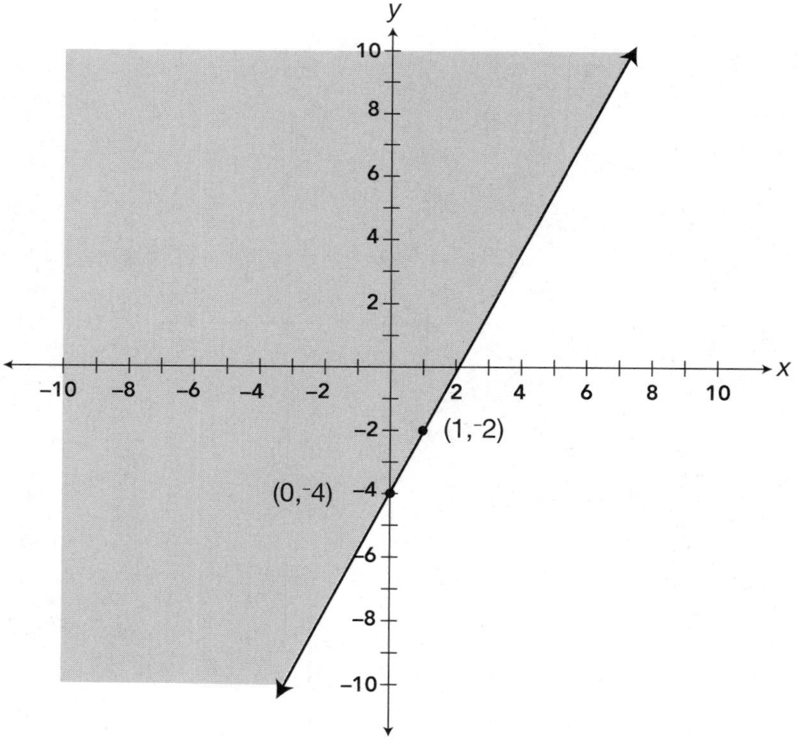

241. Multiply both sides of the inequality by 3. $3(\frac{y}{3}) < 3(\frac{2}{3} - x)$
Use the distributive property of multiplication. $3(\frac{y}{3}) < 3(\frac{2}{3}) - 3(x)$
Simplify terms. $y < 2 - 3x$
Use the commutative property of addition. $y < {}^{-}3x + 2$
The inequality is now in the proper slope/*y*-intercept form.

slope: $m = \frac{\text{change in } y}{\text{change in } x} = {}^{-}3 = \frac{{}^{-}3}{1}$

y-intercept: $b = 2$. The *y*-intercept is the point (0,2).

Point on the line: Starting at (0,2), a change in *y* of $^{-}3$ and in *x* of 1 gives the point (0+1,2−3) or (1,$^{-}$1).

Draw a dotted boundary line and shade below it.

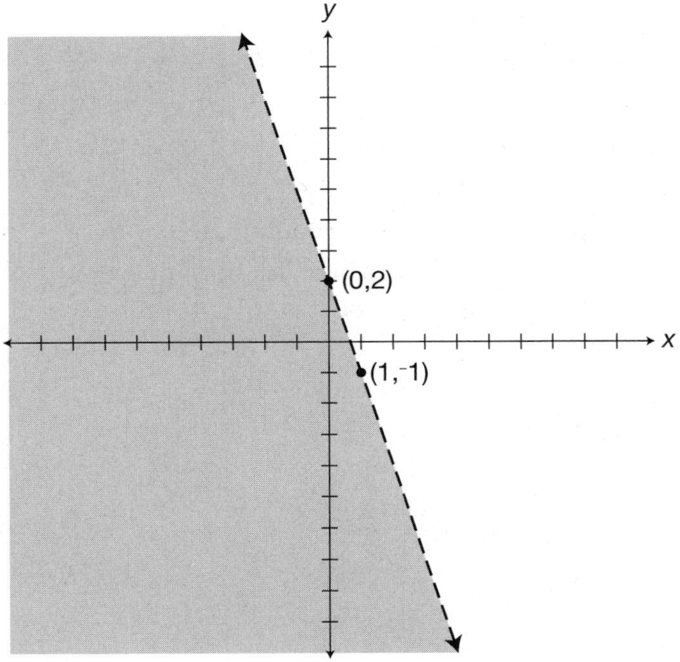

242. Put the inequality in the proper slope/y-intercept form.

Multiply both sides of the inequality by 9 (to clear the fractions).

$$9(^-\tfrac{2}{9}y + \tfrac{2}{3}x) \leq 9(^-2)$$

$$9(^-\tfrac{2}{9}y) + 9(\tfrac{2}{3}x) \leq 9(^-2)$$

$$^-2y + 6x \leq {^-18}$$

Subtract $6x$ from both sides of the inequality.

$$^-2y + 6x - 6x \leq {^-18} - 6x$$

Combine like terms on the left side of the inequality.

$$^-2y \leq {^-18} - 6x$$

Divide both sides of the inequality by -2. (Remember to switch the inequality sign because you are dividing by a negative number.)

$$\frac{^-2y}{^-2} \geq \frac{^-18 - 6x}{^-2}$$

Simplify the inequality.

$$y \geq \frac{^-18}{^-2} + \frac{^-6x}{^-2}$$

$$y \geq 9 + 3x$$

$$y \geq 3x + 9$$

<u>slope:</u> $m = \frac{\text{change in } y}{\text{change in } x} = 3 = \frac{3}{1}$

<u>y-intercept:</u> $b = 9$, so the y-intercept is the point (0,9).

<u>Point on the line:</u> Starting at (0,9), a change in y of 3 and in x of 1 gives the point (0+1,9+3) = (1,12).

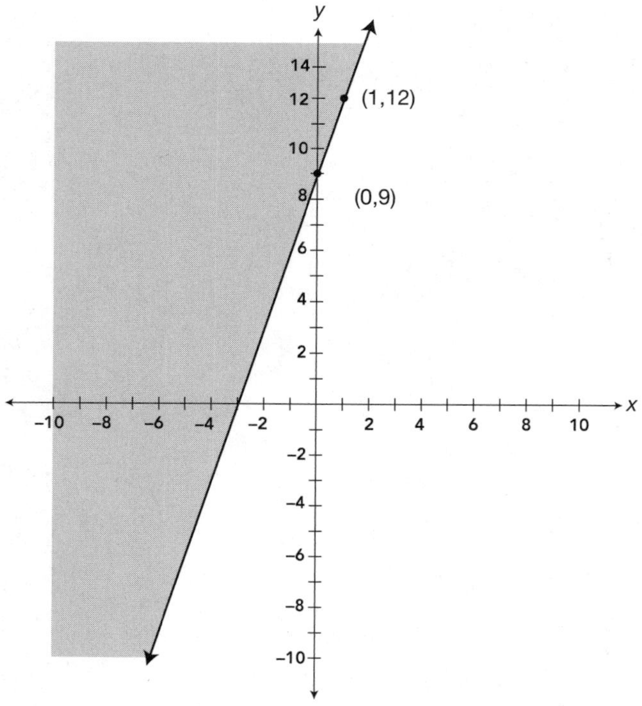

243. Subtract $0.3y$ from both sides of the inequality. $0.5x - 0.3y > 0.3y - 0.3y - 0.9$

Combine like terms. $0.5x - 0.3y > {}^-0.9$

Subtract $0.5x$ from both sides of the inequality. $0.5x - 0.5x - 0.3y > {}^-0.5x - 0.9$

Combine like terms. ${}^-0.3y > {}^-0.5x - 0.9$

Divide both sides of the inequality by $^-0.3$ and change the direction of the inequality symbol. $\frac{^-0.3y}{^-0.3} < \frac{^-0.5x - 0.9}{^-0.3}$

Simplify the expression. $\frac{^-0.3y}{^-0.3} < (\frac{^-0.5x}{^-0.3}) - (\frac{0.9}{^-0.3})$

Simplify the terms. $y < \frac{5}{3}x - (^-3)$

Subtracting a negative number is the same as adding a positive. $y < \frac{5}{3}x + 3$

The inequality is now in the proper slope/y-intercept form.

<u>slope:</u> $m = \frac{\text{change in } y}{\text{change in } x} = \frac{5}{3}$

<u>y-intercept:</u> $b = 3$. The y-intercept is the point (0,3).

<u>Point on the line:</u> Starting at (0,3), a change in y of 5 and in x of 3 gives the point (0+3,3+5) or (3,8).

Draw a dotted boundary line and shade below it.

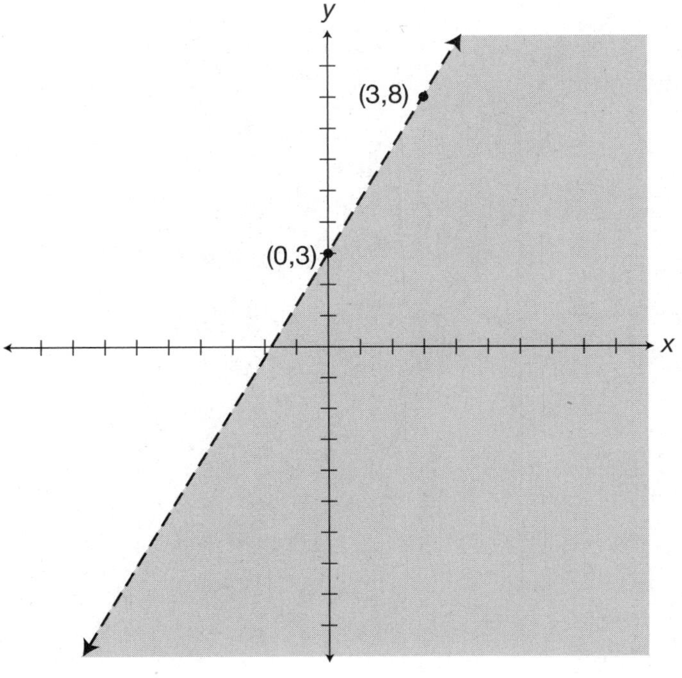

244. Subtract $3x$ from both sides of the inequality. $3x - 3x - y \leq 7x - 3x + y - 8$

Combine like terms. $^-y \leq 4x + y - 8$

Subtract y from both sides of the inequality. $^-y - y \leq 4x + y - y - 8$

Combine like terms. $^-2y \leq 4x - 8$

Divide both sides of the inequality by $^-2$ and change the direction of the inequality symbol. $\frac{^-2y}{^-2} \geq \frac{4x - 8}{^-2}$

$y \geq \frac{4x}{^-2} - (\frac{8}{^-2})$

Simplify terms. $y \geq {}^-2x - ({}^-4)$

Simplify. $y \geq {}^-2x + 4$

The inequality is now in the proper slope/y-intercept form.

slope: $m = \frac{\text{change in } y}{\text{change in } x} = {}^-2 = \frac{^-2}{1}$

y-intercept: $b = 4$. The y-intercept is the point (0,4).

Point on the line: Starting at (0,4), a change in y of $^-2$ and in x of 1 gives the point (0+1,4–2) or (1,2).

Draw a solid boundary line and shade above it.

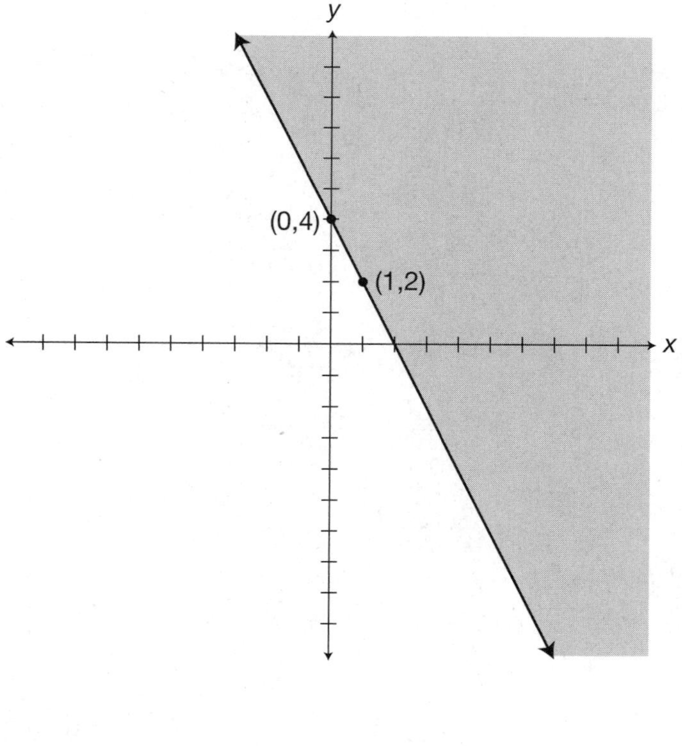

245. Put the inequality in the proper slope/y-intercept form.

$$^-(2) - (^-3y) < 1 + 3(2) + 3(^-x)$$

Use the distributive law of multiplication on both sides of the inequality.

$$^-2 + 3y < 1 + 6 - 3x$$
$$^-2 + 3y < 7 - 3x$$

Add 2 to both sides of the inequality.

$$^-2 + 2 + 3y < 7 + 2 - 3x$$

Combine like terms on the both sides of the inequality.

$$3y < 9 - 3x$$

Divide both sides of the inequality by 3.

$$\frac{3y}{3} < \frac{9-3x}{3}$$
$$y < \frac{9}{3} + \frac{-3x}{3}$$

Simplify the inequality.

$$y < 3 - x$$
$$y < {}^-x + 3$$

<u>slope:</u> $m = \frac{\text{change in } y}{\text{change in } x} = {}^-1 = \frac{^-1}{1}$

<u>y-intercept:</u> $b = 3$, so the y-intercept is the point (0,3).

<u>Point on the line:</u> Starting at (0,3), a change in y of $^-1$ and in x of 1 gives the point $(0+1,3-1) = (1,2)$.

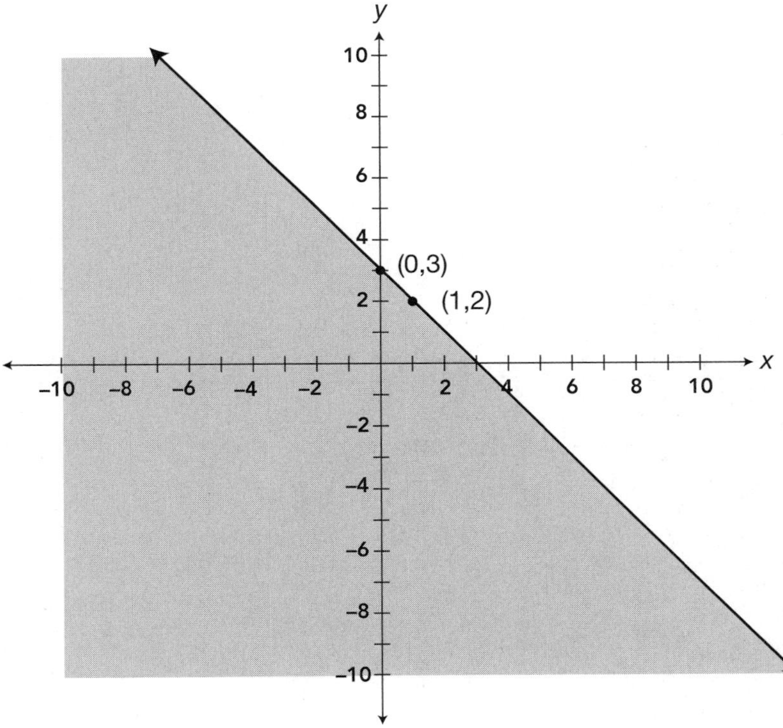

246. Use the distributive property of multiplication. $^-12 \leq {}^-3x - 3y$

Add $3y$ to both sides of the inequality. $3y - 12 \leq {}^-3x - 3y + 3y$

Combine like terms. $3y - 12 \leq {}^-3x$

Add 12 to both sides of the inequality. $3y - 12 + 12 \leq {}^-3x + 12$

Combine like terms. $3y \leq {}^-3x + 12$

Divide both sides of the inequality by 3. $\frac{3y}{3} \leq \frac{^-3x}{3} + \frac{12}{3}$

Simplify the expressions. $y \leq {}^-x + 4$

The inequality is already in the proper slope/y-intercept form.

<u>slope:</u> $m = \frac{\text{change in } y}{\text{change in } x} = {}^-1 = \frac{^-1}{1}$

<u>y-intercept:</u> $b = 4$. The y-intercept is the point (0,4).

<u>Point on the line:</u> Starting at (0,4), a change in y of $^-1$ and in x of 1 gives the point (0+1,4–1) or (1,3).

Draw a solid boundary line and shade below it.

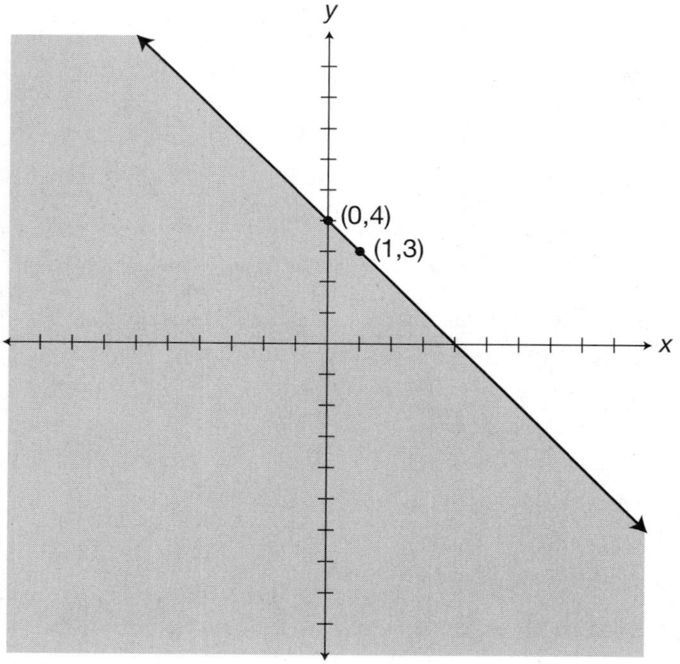

247. Use the distributive property of multiplication. $9y + 7 \geq 2x + 16$

Subtract 7 from both sides of the inequality. $9y + 7 - 7 \geq 2x + 16 - 7$

Combine like terms. $9y \geq 2x + 9$

Divide both sides of the inequality by 9. $\frac{9y}{9} \geq \frac{2x}{9} + \frac{9}{9}$

Simplify the expressions. $y \geq \frac{2}{9}x + 1$

The inequality is now in the proper slope/y-intercept form.

<u>slope:</u> $m = \frac{\text{change in } y}{\text{change in } x} = = \frac{2}{9}$

<u>y-intercept:</u> $b = 1$. The y-intercept is the point (0,1).

<u>Point on the line:</u> Starting at (0,1), a change in y of 2 and in x of 9 gives the point (0+9,1+2) or (9,3).

Draw a solid boundary line and shade above it.

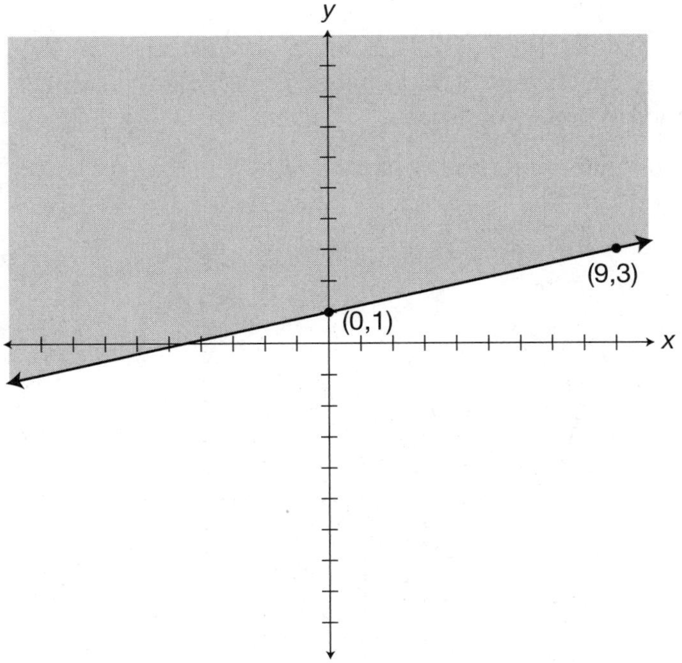

248. Use the distributive property of multiplication. $2(y) + 2(3) - x \geq 6(1) - 6(x)$

Simplify terms. $2y + 6 - x \geq 6 - 6x$

Add x to both sides of the inequality. $2y + 6 - x + x \geq 6 - 6x + x$

Combine like terms. $2y + 6 \geq 6 - 5x$

Subtract 6 from both sides of the inequality. $2y + 6 - 6 \geq 6 - 5x - 6$

Use the commutative property with like terms. $2y + 6 - 6 \geq 6 - 6 - 5x$

Combine like terms. $2y \geq {}^-5x$

Divide both sides of the inequality by 2. $\frac{2y}{2} \geq \frac{{}^-5x}{2}$

Simplify terms. $y \geq \frac{{}^-5}{2}x$

The inequality is already in the proper slope/y-intercept form.

<u>slope:</u> $m = \frac{\text{change in } y}{\text{change in } x} = \frac{{}^-5}{2}$

<u>y-intercept:</u> $b = 0$. The y-intercept is the point (0,0).

<u>Point on the line:</u> Starting at (0,0), a change in y of $^-5$ and in x of 2 gives the point (0+2,0−5) or (2,$^-$5).

Draw a solid boundary line and shade above it.

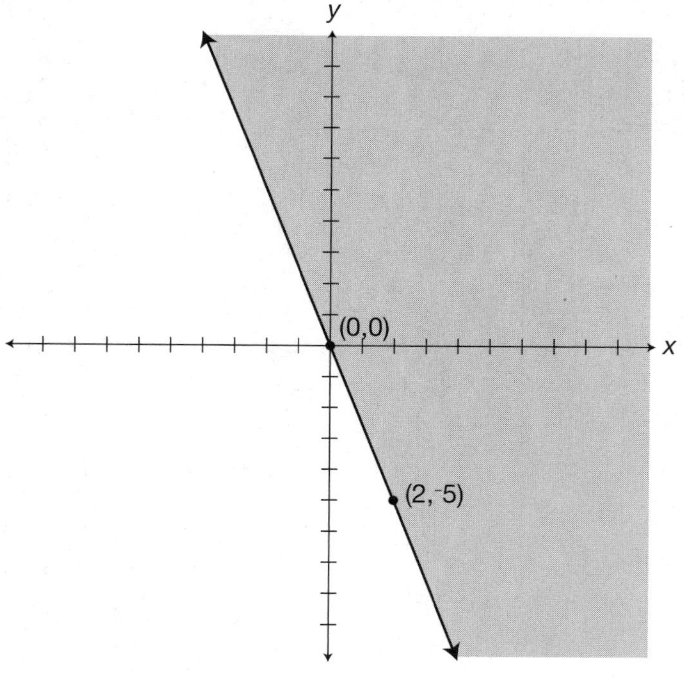

249. Use the distributive property of multiplication. $^-28y \geq 2x - 14(y) - 14(10)$

Simplify terms. $^-28y \geq 2x - 14y - 140$

Add 14*y* to both sides of the inequality. $^-28y + 14y \geq 2x - 14y + 14y - 140$

Combine like terms on each side of the inequality. $^-14y \geq 2x - 140$

Divide both sides of the inequality by $^-14$ and change the direction of the inequality symbol. $\frac{^-14y}{^-14} \leq \frac{2x}{^-14} - \frac{140}{^-14}$

Simplify the terms. $y \leq \frac{^-1}{7}x + 10$

The inequality is now in the proper slope/*y*-intercept form.

slope: $m = \frac{\text{change in } y}{\text{change in } x} = \frac{^-1}{7}$

y-intercept: $b = 10$. The *y*-intercept is the point (0,10).

Point on the line: Starting at (0,10), a change in *y* of $^-1$ and in *x* of 7 gives the point (0+7,10−1) or (7,9).

Draw a solid boundary line and shade below it.

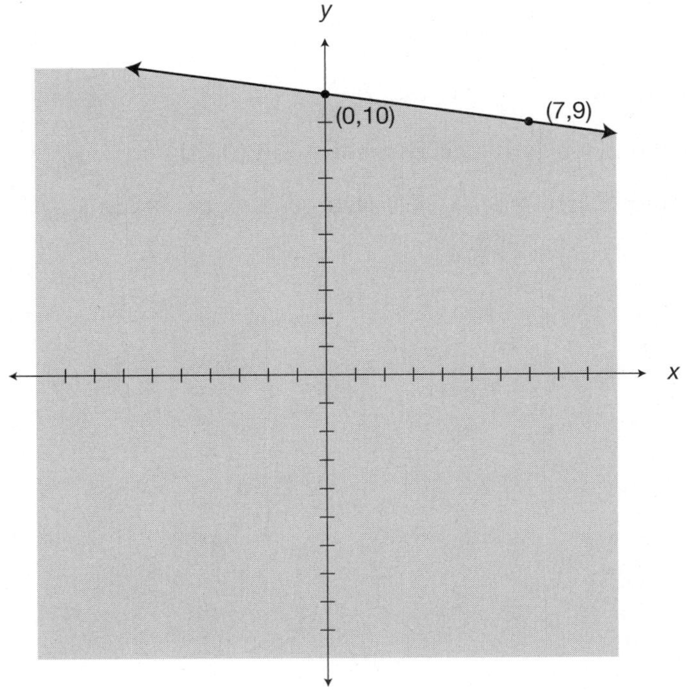

250. Put the inequality in the proper slope/y-intercept form.

Divide both sides of the inequality by -2. (Remember to reverse the inequality sign because you are dividing by a negative number.) $1 - 2(2y - 3x) \geq 0$

Use the distributive law of multiplication to simplify the left side of the inequality.

$1 - 2(2y) - 2(^{-}3x) \geq 0$

$1 - 4y + 6x \geq 0$

Add $4y$ to both sides of the inequality. $1 - 4y + 4y + 6x \geq 0 + 4y$

Combine like terms on the left side of the inequality. $1 + 6x \geq 4y$

Divide both sides of the inequality by 4. $\frac{1+6x}{4} \geq \frac{4y}{4}$

Simplify the inequality.

$\frac{1+6x}{4} \geq y$

$\frac{1}{4} + \frac{6x}{4} \geq y$

$\frac{1}{4} + \frac{3x}{2} \geq y$

If $b \leq a$, then $a \geq b$.

slope: $m = \frac{\text{change in } y}{\text{change in } x} = \frac{3}{2}$

y-intercept: $b = \frac{1}{4}$, so the y-intercept is the point $(0,\frac{1}{4})$.

Point on the line: Starting at $(0,\frac{1}{4})$, a change in y of 3 and in x of 2 gives the point $(0+2,\frac{1}{4}+3)=\left(2,\frac{13}{4}\right)$

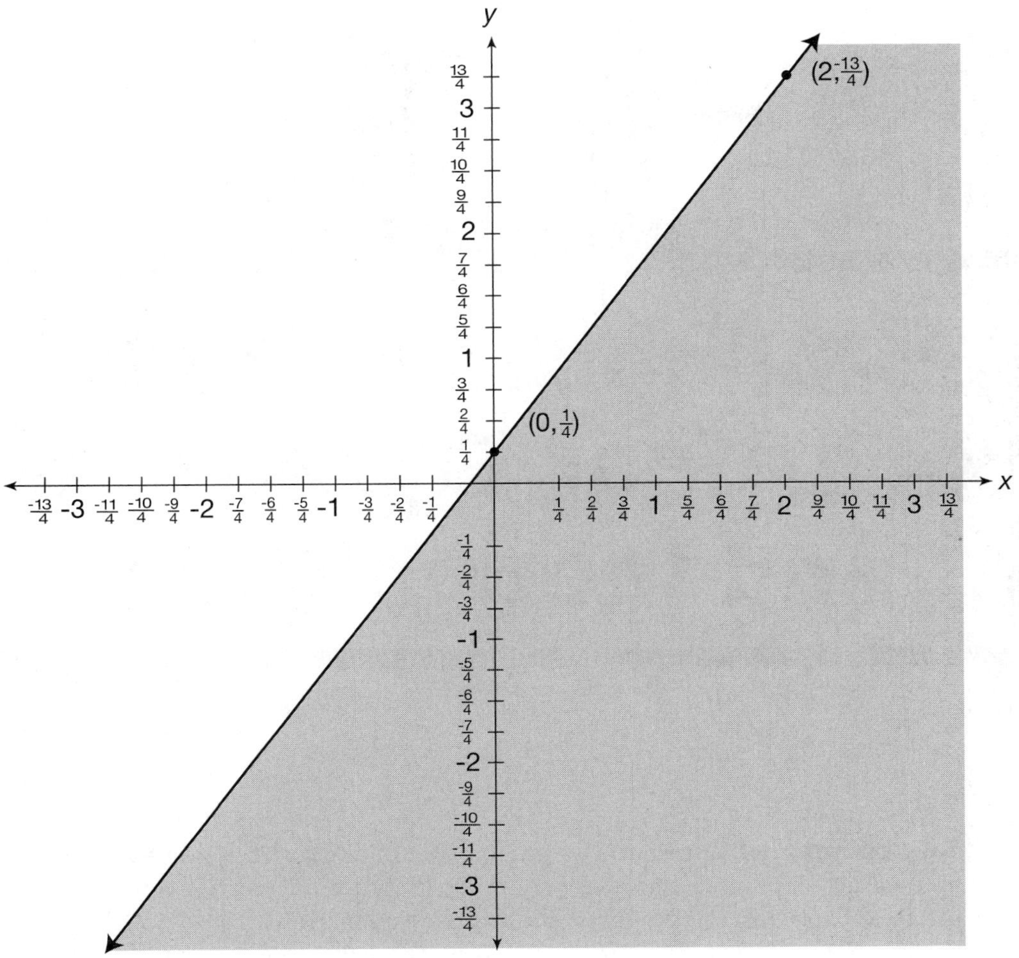

Graph for question 250

11

Graphing Systems of Linear Equations and Inequalities

This chapter will present 15 systems of equalities and ten systems of inequalities as practice in finding solutions graphically. You will find complete explanations and graphs in the answer explanations.

Graphing systems of linear equations on the same coordinate plane will give you solutions common to both equations. There are three possibilities for a pair of equations:

- The solution will be one coordinate pair at the point of intersection.
- The solution will be all the points on the line graph because the equations coincide.
- There will be no solution if the line graphs have the same slope but different y-intercepts. In this case, the lines are parallel and will not intersect.

Pairs of inequalities can also have a common solution. The graphic solution will either be the common areas of the graphs of the inequalities or there will be no solution if the areas do not overlap.

Tips for Graphing Systems of Linear Equations and Inequalities

Transform each equation or inequality into the slope/y-intercept form.

For equations, graph the lines and look for the point or points of intersection. That is the solution.

For inequalities, graph the boundaries as the appropriate dotted or solid line and shade the area for each inequality depending upon the inequality symbol present. The intersection of the shaded areas will be the solution for the system.

When multiplying or dividing by a negative term, change the direction of the inequality symbol for each operation.

Find the solutions for the following systems of equations by graphing on graph paper.

251. $y = x + 4$
$y = {}^-x + 2$

252. $2y - x = 2$
$3x + y = 8$

253. $4y = {}^-7(x + 4)$
$4y = x + 4$

254. $y - x = 5 - x$
${}^-4y = 8 - 7x$

255. $2y = 6x + 14$
$4y = x - 16$

256. $2x + y = 4$
$3(y + 9) = 7x$

257. $y = x + 9$
$4y = 16 - x$

258. $4x - 5y = 5$
$5y = 20 - x$

259. $6y = 9(x - 6)$
$3(2y + 5x) = {}^-6$

260. $15y = 6(3x + 15)$
$y = 6(1 - x)$

261. $3y = 6x + 6$
$5y = 10(x - 5)$

262. $3(2x + 3y) = 63$
$27y = 9(x - 6)$

263. $x - 20 = 5y$
$10y = 8x + 20$

264. $3x + 4y = 12$
$y = 3 - \frac{3}{4}x$

265. $16y = 10(x - 8)$
$8y - 17x = 56$

Find the solution for each of the following systems of inequalities by graphing and shading.

266. $2y - 3x \geq {}^{-}6$
$y \geq 5 - \frac{5}{2}x$

267. $6y < 5x - 30$
$2y < {}^{-}x + 4$

268. $y - x \geq 6$
$11y \geq {}^{-}2(x + 11)$

269. $5y \leq 8(x + 5)$
$5y < 12(5 - x)$

270. $2(x + 5y) > 5(x + 6)$
$4x + y < 4x + 5$

271. $3y \leq {}^{-}3(x + 3)$
$3y \geq 2(6 - x)$

272. $9(y - 4) < 4x$
${}^{-}9y < 2(x + 9)$

273. $7(y - 5) < {}^{-}5x$
${}^{-}3 < \frac{1}{4}(2x - 3y)$

274. $y > \frac{7}{4}(4 - x)$
$3(y + 5) > 7x$

275. $5x - 2(y + 10) \leq 0$
$2x + y \leq {}^{-}3$

Answers

Numerical expressions in parentheses like this [ ] are operations performed on only part of the original expression. The operations performed within these symbols are intended to show how to evaluate the various terms that make up the entire expression.

Expressions with parentheses that look like this ( ) contain either numerical substitutions or expressions that are part of a numerical expression. Once a single number appears within these parentheses, the parentheses are no longer needed and need not be used the next time the entire expression is written.

When two pair of parentheses appear side by side like this ( )( ), it means that the expressions within are to be multiplied.

Sometimes parentheses appear within other parentheses in numerical or algebraic expressions. Regardless of what symbol is used, ( ), { }, or [ ], perform operations in the innermost parentheses first and work outward.

The solution is described verbally in each case and is underlined. The graph is shown.

251. Transform equations into slope/y-intercept form. $y = x + 4$
The equation is in the proper slope/y-intercept form.

slope: $m = \frac{\text{change in } y}{\text{change in } x} = 1 = \frac{1}{1}$

y-intercept: $b = 4$. The y-intercept is the point (0,4).
Starting at (0,4), the slope tells you to go up 1 and right 1 for (1,5).

* * * * * * * * * * * *

The equation is already in the proper slope/y-intercept form. $y = {}^{-}x + 2$

slope: $m = \frac{\text{change in } y}{\text{change in } x} = {}^{-}1 = \frac{{}^{-}1}{1}$

y-intercept: $b = 2$. The y-intercept is the point (0,2).
Starting at (0,2), the slope tells you to go down 1 and right 1 for (1,1).

The solution of the system is (⁻1,3).

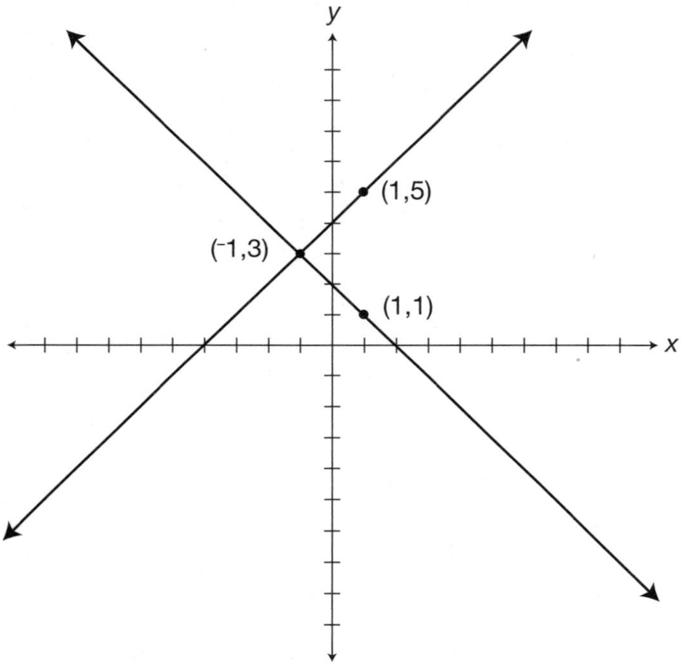

252. Transform equations into slope/y-intercept form. $2y - x = 2$

Add x to both sides. $2y - x + x = x + 2$

Combine like terms. $2y = x + 2$

Divide both sides by 2. $y = \frac{1}{2}x + 1$

The equation is now in the proper slope/y-intercept form.

slope: $m = \dfrac{\text{change in } y}{\text{change in } x} = \frac{1}{2}$

y-intercept: $b = 1$. The y-intercept is the point (0,1).

Starting at (0,1), the slope tells you to go up 1 and right 2 for (2,2).

* * * * * * * * * * * *

$3x + y = 8$

Subtract $3x$ from both sides. $3x - 3x + y = {}^{-}3x + 8$

Simplify. $y = {}^{-}3x + 8$

The equation is now in the proper slope/y-intercept form.

slope: $m = \dfrac{\text{change in } y}{\text{change in } x} = {}^{-}3 = \frac{{}^{-}3}{1}$

y-intercept: $b = 8$. The y-intercept is the point (0,8).

Starting at (0,8), the slope tells you to go down 3 and right 1 for (1,5).

The solution of the system is (2,2).

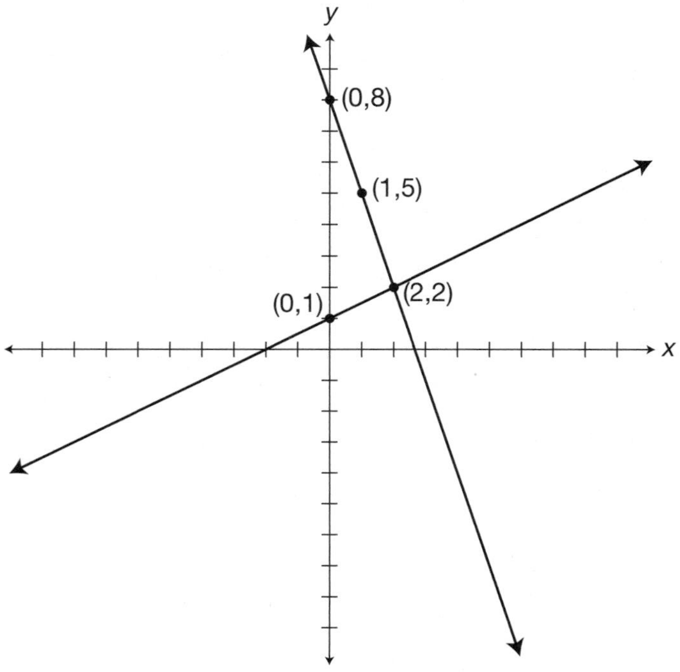

253. Transform equations into slope/y-intercept form. $4y = {}^{-}7(x + 4)$
Use the distributive property of multiplication. $4y = {}^{-}7x - 28$

Divide both sides by 4. $y = \frac{-7}{4}x - 7$
The equation is now in the proper slope/y-intercept form.

slope: $m = \frac{\text{change in } y}{\text{change in } x} = \frac{-7}{4}$

y-intercept: $b = {}^{-}7$. The y-intercept is the point $(0,{}^{-}7)$.
Starting at $(0,{}^{-}7)$, the slope tells you to go down 7 and right 4 for $(4,{}^{-}14)$.
* * * * * * * * * * * *

$4y = x + 4$

Divide both sides by 4. $y = \frac{1}{4}x + 1$
The equation is now in the proper slope/y-intercept form.

slope: $m = \frac{\text{change in } y}{\text{change in } x} = \frac{1}{4}$

y-intercept: $b = 1$. The y-intercept is the point $(0,1)$.
Starting at $(0,1)$, the slope tells you to go up 1 and right 4 for $(4,2)$.

The solution of the system is $({}^{-}4,0)$.

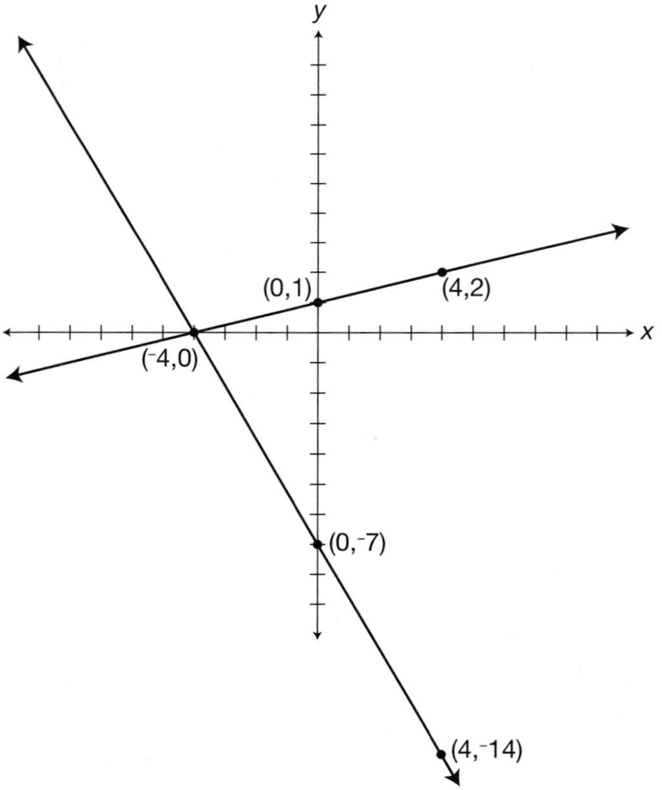

254. Transform equations into slope/y-intercept form. $y - x = 5 - x$
Add x to both sides. $y - x + x = 5 - x + x$
Combine like terms on each side. $y = 5$
The graph is a line parallel to the x-axis through (0,5).
Note the slope is 0 and so, there is no change in y as the x changes.

* * * * * * * * * * * *

$^-4y = 8 - 7x$

Divide both sides by $^-4$. $\frac{^-4y}{^-4} = \frac{8}{^-4} - \frac{7x}{^-4}$

Simplify terms. $y = {}^-2 + \frac{7}{4}x$

Use the commutative property. $y = \frac{7}{4}x - 2$
The equation is now in the proper slope/y-intercept form.

slope: $m = \frac{\text{change in } y}{\text{change in } x} = \frac{7}{4}$

y-intercept: $b = {}^-2$. The y-intercept is the point $(0,^-2)$.
Starting at $(0,^-2)$, the slope tells you to go up 7 and right 4 for (4,5).

The solution of the system is (4,5).

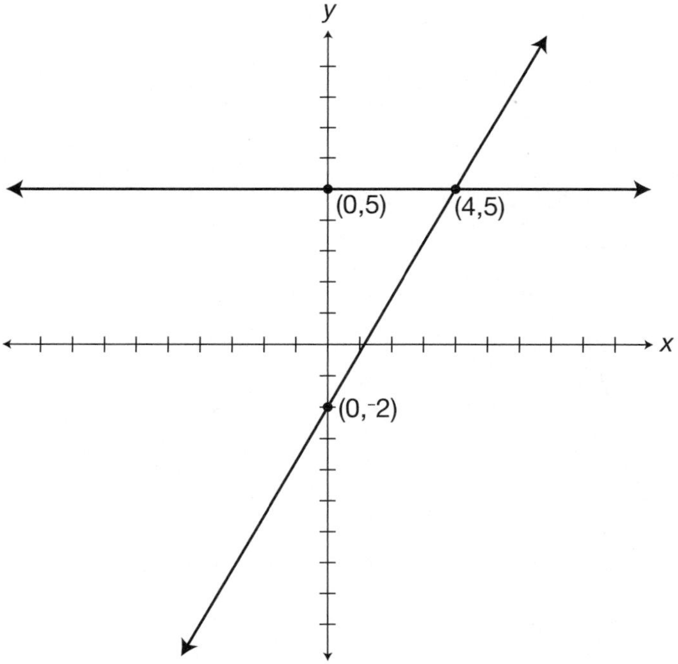

255. Transform equations into slope/y-intercept form.

Divide both sides by 2.

$2y = 6x + 14$

$y = \frac{6}{2}x + \frac{14}{2}$

$y = 3x + 7$

The equation is now in the proper slope/y-intercept form.
Use the negatives to keep the coordinates near the origin.

slope: $m = \frac{\text{change in } y}{\text{change in } x} = 3 = \frac{3}{1} = \frac{^-3}{^-1}$

y-intercept: $b = 7$. The y-intercept is the point (0,7).
Starting at (0,7), the slope tells you to go down 3 and left 1 for ($^-1$,4).

* * * * * * * * * * * *

$4y = x - 16$

Divide both sides by 4.

$y = \frac{1}{4}x - 4$

The equation is now in the proper slope/y-intercept form.

slope: $m = \frac{\text{change in } y}{\text{change in } x} = \frac{1}{4}$

y-intercept: $b = {}^-4$. The y-intercept is the point (0,$^-4$).
Starting at (0,$^-4$), the slope tells you to go up 1 and right 4 for (4,$^-3$).

The solution of the system is ($^-4$,$^-5$).

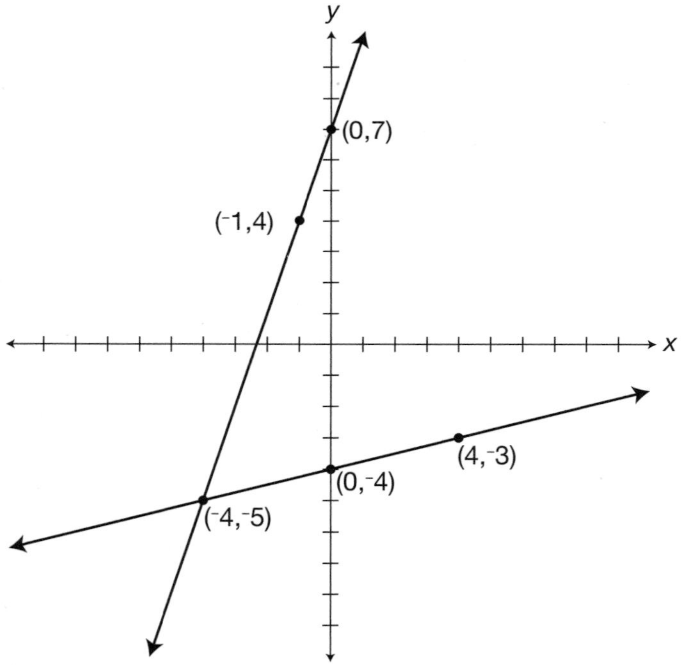

256. Transform equations into slope/y-intercept form. $2x + y = 4$
Subtract $2x$ from both sides. $2x - 2x + y = 4 - 2x$
Combine like terms on each side. $y = 4 - 2x$
Use the commutative property. $y = {}^-2x + 4$
The equation is now the proper slope/y-intercept form.

slope: $m = \frac{\text{change in } y}{\text{change in } x} = {}^-2 = \frac{^-2}{1}$

y-intercept: $b = 4$. The y-intercept is the point (0,4).
Starting at (0,4), the slope tells you to go down 2 and right 1 for (1,2).

* * * * * * * * * * * *

$3(y + 9) = 7x$
Use the distributive property of multiplication. $3y + 27 = 7x$
Subtract 27 from both sides. $3y + 27 - 27 = 7x - 27$
Simplify. $3y = 7x - 27$
Divide both sides by 3. $y = \frac{7}{3}x - 9$
The equation is now in the proper slope/y-intercept form.

slope: $m = \frac{\text{change in } y}{\text{change in } x} = \frac{7}{3}$

y-intercept: $b = {}^-9$. The y-intercept is the point (0,$^-$9).
Starting at (0,$^-$9), the slope tells you to go up 7 and right 3 for (3,$^-$2).

The solution of the system is (3,$^-$2).

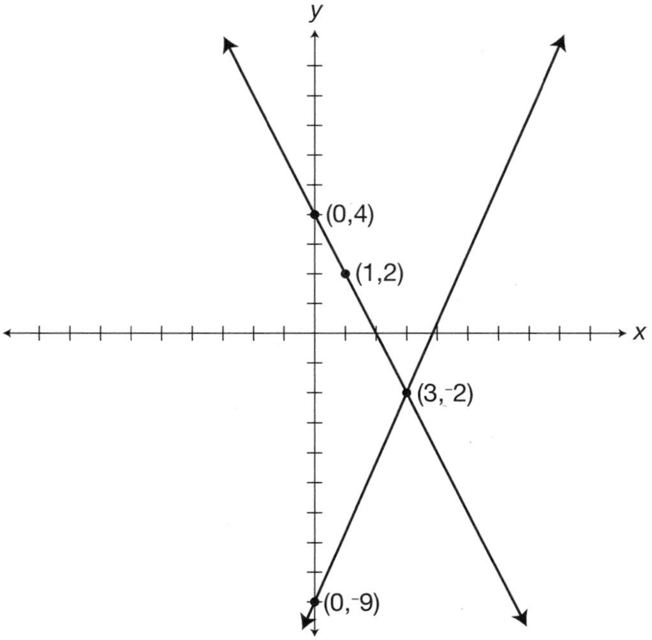

257. Transform equations into slope/y-intercept form. $y = x + 9$

The equation is already in the proper slope/y-intercept form.

slope: $m = \frac{\text{change in } y}{\text{change in } x} = 1 = \frac{1}{1}$

y-intercept: $b = 9$. The y-intercept is the point (0,9).

Starting at (0,9), the slope tells you to go up 1 and right 1 for (1,10).

* * * * * * * * * * * *

$4y = 16 - x$

Use the commutative property. $4y = {}^{-}x + 16$

Divide both sides by 4. $y = -\frac{1}{4}x + 4$

The equation is now in the proper slope/y-intercept form

slope: $m = \frac{\text{change in } y}{\text{change in } x} = \frac{-1}{4}$

y-intercept: $b = 4$. The y-intercept is the point (0,4).

Starting at (0,4), the slope tells you to go down 1 and right 4 for (4,3).

The solution of the system is (⁻4,5).

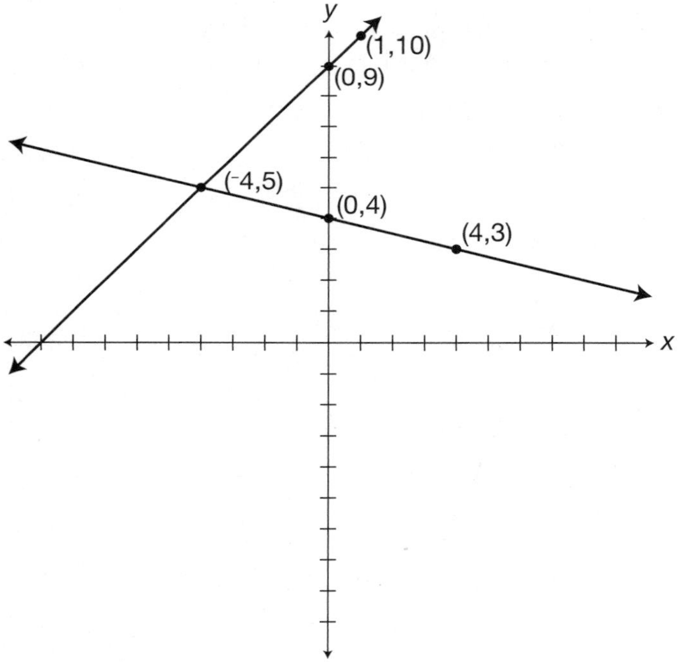

258. Transform equations into slope/y-intercept form. $4x - 5y = 5$

Subtract $4x$ from both sides. $4x - 4x - 5y = 5 - 4x$

Simplify. $^-5y = 5 - 4x$

Use the commutative property. $^-5y = {}^-4x + 5$

Divide both sides by $^-5$. $y = \frac{4}{5}x - 1$

The equation is now in the proper slope/y-intercept form.

<u>slope:</u> $m = \frac{\text{change in } y}{\text{change in } x} = \frac{4}{5}$

<u>y-intercept:</u> $b = {}^-1$. The y-intercept is the point $(0,{}^-1)$.

Starting at $(0,{}^-1)$, the slope tells you to go up 4 and right 5 for (5,3).

* * * * * * * * * * * *

$5y = 20 - x$

Divide both sides by 5. $y = 4 - \frac{1}{5}x$

Use the commutative property. $y = \frac{-1}{5}x + 4$

The equation is now in the proper slope/y-intercept form.

<u>slope:</u> $m = \frac{\text{change in } y}{\text{change in } x} = \frac{-1}{5}$

<u>y-intercept:</u> $b = 4$. The y-intercept is the point (0,4).

Starting at (0,4), the slope tells you to go down 1 space and right 5 for (5,3).

<u>The solution of the system is (5,3).</u>

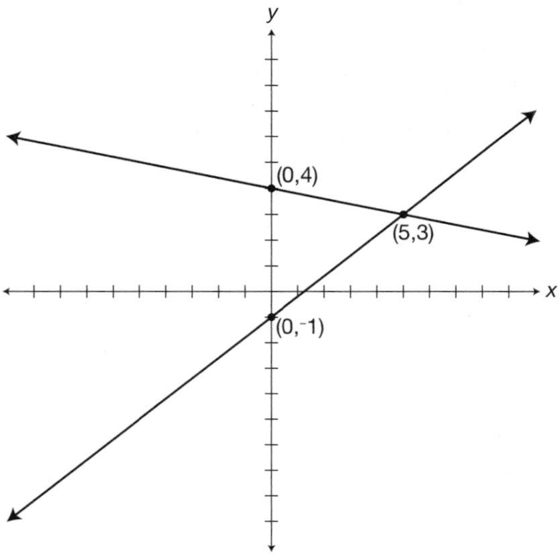

259. Transform equations into slope/y-intercept form. $6y = 9(x - 6)$

Use the distributive property of multiplication. $6y = 9x - 54$

Divide both sides by 6. $y = \frac{9}{6}x - 9$

Simplify. $y = \frac{3}{2}x - 9$

The equation is now in the proper slope/y-intercept form.

slope: $m = \frac{\text{change in } y}{\text{change in } x} = \frac{3}{2}$

y-intercept: $b = {}^-9$. The y-intercept is the point $(0,{}^-9)$.

Starting at $(0,{}^-9)$, the slope tells you to go up 3 spaces and right 2 for $(2,{}^-6)$.

* * * * * * * * * * * *

$3(2y + 5x) = {}^-6$

Use the distributive property of multiplication. $6y + 15x = {}^-6$

Subtract $15x$ from both sides. $6y + 15x - 15x = {}^-15x - 6$

Simplify. $6y = {}^-15x - 6$

Divide both sides by 6. $y = \frac{{}^-15}{6}x - 1$

Simplify. $y = -\frac{5}{2}x - 1$

The equation is now in the proper slope/y-intercept form.

slope: $m = \frac{\text{change in } y}{\text{change in } x} = \frac{{}^-5}{2}$

y-intercept: $b = {}^-1$. The y-intercept is the point $(0,{}^-1)$.

Starting at $(0,{}^-1)$, the slope tells you to go down 5 and right 2 for $(2,{}^-6)$.

The solution of the system is $(2,{}^-6)$.

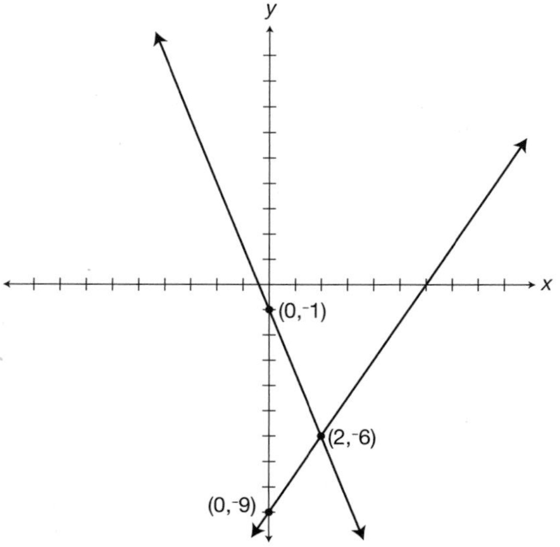

260. Transform equations into slope/y-intercept form. $15y = 6(3x + 15)$

Use the distributive property of multiplication. $15y = 18x + 90$

Divide both sides by 15. $y = \frac{18}{15}x + 6$

Simplify. $y = \frac{6}{5}x + 6$

The equation is now in the proper slope/y-intercept form.

slope: $m = \frac{\text{change in } y}{\text{change in } x} = \frac{6}{5} = \frac{^-6}{^-5}$

y-intercept: $b = 6$. The y-intercept is the point (0,6).

Starting at (0,6), the slope tells you to go down 6 and left 5 for ($^-5$,0).

* * * * * * * * * * * *

$y = 6(1 - x)$

Use the distributive property of multiplication. $y = 6 - 6x$

Use the commutative property. $y = {}^-6x + 6$

The equation is now in the proper slope/y-intercept form.

slope: $m = \frac{\text{change in } y}{\text{change in } x} = {}^-6 = \frac{^-6}{1}$

y-intercept: $b = 6$. The y-intercept is the point (0,6).

Starting at (0,6), the slope tells you to go down 6 and right 1 for (1,0).

The solution of the system is (0,6).

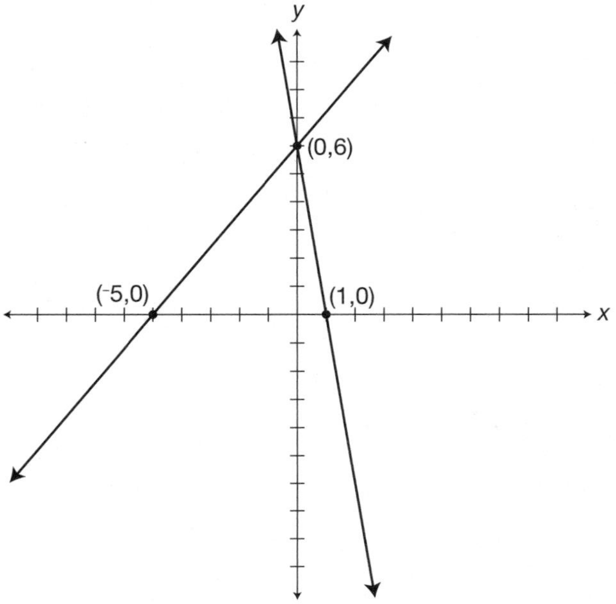

261. Transform equations into slope/y-intercept form. $3y = 6x + 6$

Divide both sides by 3. $y = 2x + 2$

The equation is now in the proper slope/y-intercept form.

slope: $m = \frac{\text{change in } y}{\text{change in } x} = 2 = \frac{2}{1}$

y-intercept: $b = 2$. The y-intercept is the point (0,2).

Starting at (0,2), the slope tells you to go up 2 and right 1 for (1,4).

* * * * * * * * * * * *

$5y = 10(x - 5)$

Use the distributive property. $5y = 10x - 50$

Divide both sides by 5. $y = 2x - 10$

The equation is now in the proper slope/y-intercept form.

slope: $m = \frac{\text{change in } y}{\text{change in } x} = 2 = \frac{2}{1}$

y-intercept: $b = {}^-10$. The y-intercept is the point $(0,{}^-10)$.

Starting at $(0,{}^-10)$, the slope tells you to go up 2 and right 1 for $(1,{}^-8)$.

The lines have the same slope and different y-intercepts. So, the line graphs are parallel and do not intersect.

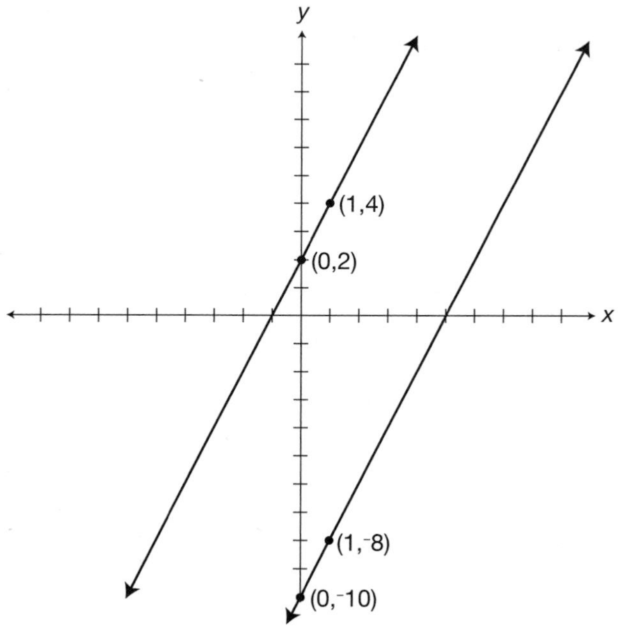

262. Transform equations into slope/y-intercept form. $3(2x + 3y) = 63$

Use the distributive property of multiplication. $6x + 9y = 63$

Subtract $6x$ from both sides. $6x - 6x + 9y = 63 - 6x$

Simplify. $9y = 63 - 6x$

Use the commutative property. $9y = {}^-6x + 63$

Divide both sides by 9. $y = {}^-\frac{6}{9}x + 7$

Simplify. $y = {}^-\frac{2}{3}x + 7$

The equation is now in the proper slope/y-intercept form.

<u>slope:</u> $m = \frac{\text{change in } y}{\text{change in } x} = {}^-\frac{2}{3}$

<u>y-intercept:</u> $b = 7$. The y-intercept is the point (0,7).

Starting at (0,7), the slope tells you to go down 2 and right 3 for (3,5).

* * * * * * * * * * * *

$27y = 9(x - 6)$

Use the distributive property of multiplication. $27y = 9x - 54$

Divide both sides by 27. $y = \frac{9}{27}x - 2$

Simplify. $y = \frac{1}{3}x - 2$

The equation is now in the proper slope/y-intercept form.

<u>slope:</u> $m = \frac{\text{change in } y}{\text{change in } x} = \frac{1}{3}$

<u>y-intercept:</u> $b = {}^-2$. The y-intercept is the point (0,$^-$2).

Starting at (0,$^-$2), the slope tells you to go up 1 and right 3 for (3,$^-$1).

<u>The solution of the system is (9,1).</u>

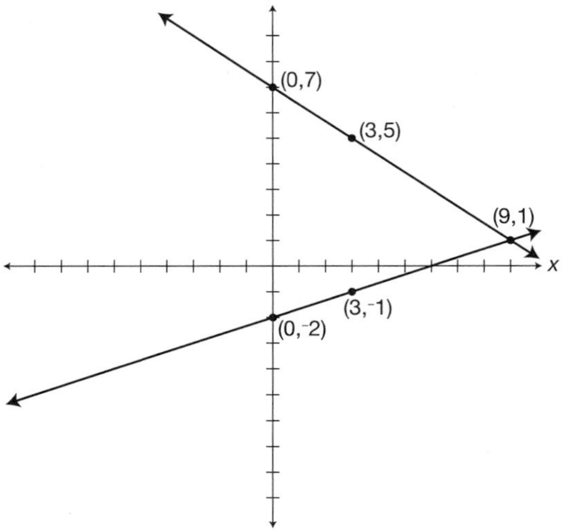

263. Transform equations into slope/y-intercept form. $x - 20 = 5y$

If $a = b$, then $b = a$. $5y = x - 20$

Divide both sides by 5. $y = \frac{1}{5}x - 4$

The equation is now in the proper slope/y-intercept form.

slope: $m = \frac{\text{change in } y}{\text{change in } x} = \frac{1}{5}$

y-intercept: $b = {}^-4$. The y-intercept is the point $(0,{}^-4)$.

Starting at $(0,{}^-4)$, the slope tells you to go up 1 and right 5 for $(5,{}^-3)$.

* * * * * * * * * * * *

$10y = 8x + 20$

Divide both sides by 10. $y = \frac{8}{10}x + 2$

Simplify. $y = \frac{4}{5}x + 2$

The equation is now in the proper slope/y-intercept form.

slope: $m = \frac{\text{change in } y}{\text{change in } x} = \frac{4}{5}$

y-intercept: $b = 2$. The y-intercept is the point $(0,2)$.

Starting at $(0,2)$, the slope tells you to go up 4 and right 5 for $(5,6)$.

The solution of the system is $({}^-10,{}^-6)$.

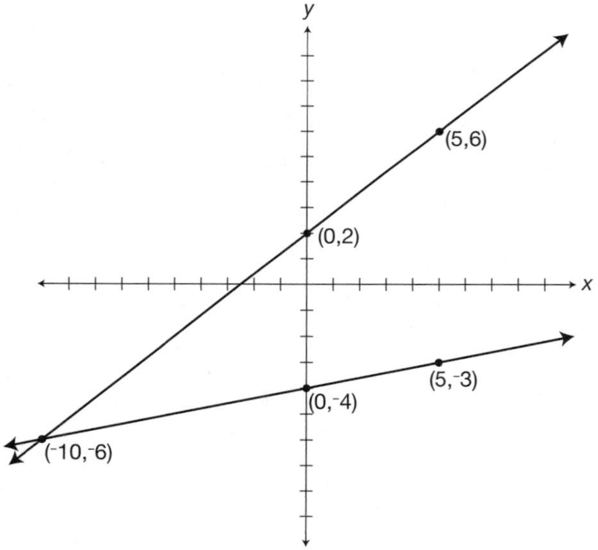

264. Transform equations into slope/y-intercept form. $3x + 4y = 12$

Subtract $3x$ from both sides. $3x - 3x + 4y = {}^{-}3x + 12$

Simplify. $4y = {}^{-}3x + 12$

Divide both sides by 4. $y = \frac{-3}{4}x + 3$

The equation is now in the proper slope/y-intercept form.

slope: $m = \frac{\text{change in } y}{\text{change in } x} = \frac{-3}{4}$

y-intercept: $b = {}^{+}3$. The y-intercept is the point (0,3).

Starting at (0,3), the slope tells you to go down 3 and right 4 for (4,0).

* * * * * * * * * * * *

$y = 3 - \frac{3}{4}x$

Use the commutative property. $y = \frac{-3}{4}x + 3$

The equation is now in the proper slope/y-intercept form.

slope: $m = \frac{\text{change in } y}{\text{change in } x} = \frac{-3}{4}$

y-intercept: $b = 3$. The y-intercept is the point (0,3).

Starting at (0,3), the slope tells you to go down 3 and right 4 for (4,0).

The solution for the system of equations is the entire line because the graphs coincide.

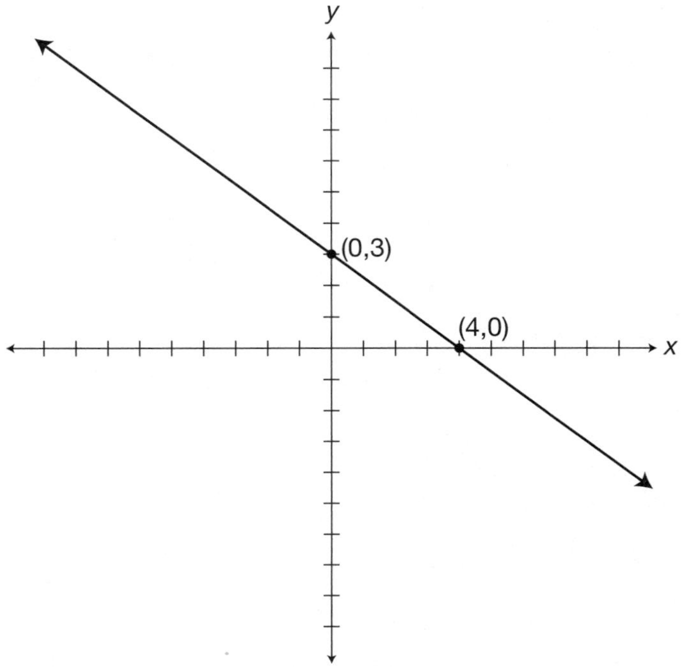

265. Transform equations into slope/y-intercept form. $16y = 10(x - 8)$

Use the distributive property of multiplication. $16y = 10x - 80$

Divide both sides by 16. $y = \frac{10}{16}x - 5$

Simplify. $y = \frac{5}{8}x - 5$

The equation is now in the proper slope/y-intercept form.

{Use the negatives to keep the coordinates near the origin.}

slope: $m = \frac{\text{change in } y}{\text{change in } x} = \frac{5}{8} = \frac{^-5}{^-8}$

y-intercept: $b = {}^-5$. The y-intercept is the point $(0,{}^-5)$.

Starting at $(0,{}^-5)$, the slope tells you to go down 5 and left 8 for $({}^-8,{}^-10)$.

* * * * * * * * * * * *

$8y - 17x = 56$

Add $17x$ from both sides. $8y - 17x + 17x = 17x + 56$

Simplify. $8y = 17x + 56$

Divide both sides by 8. $y = \frac{17}{8}x + 7$

The equation is now in the proper slope/y-intercept form.

slope: $m = \frac{\text{change in } y}{\text{change in } x} = \frac{17}{8} = \frac{^-17}{^-8}$

{Use the negatives to keep the coordinates near the origin.}

y-intercept: $b = 7$. The y-intercept is the point $(0,7)$.

Starting at $(0,7)$, the slope tells you to go down 17 and left 8 for $({}^-8,{}^-10)$.

The solution for the system of equations is $({}^-8,{}^-10)$.

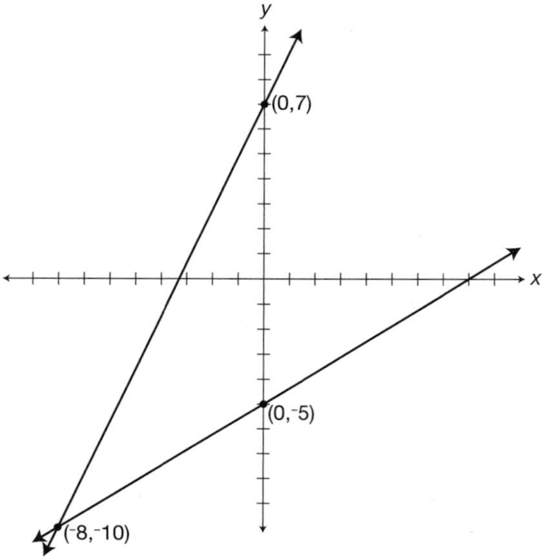

266. Transform the inequalities into slope/y-intercept form. $2y - 3x \geq {}^{-}6$

Add $3x$ to both sides. $2y - 3x + 3x \geq {}^{+}3x - 6$

Simplify. $2y \geq 3x - 6$

Divide both sides by 2. $y \geq \frac{3}{2}x - 3$

slope: $m = \frac{\text{change in } y}{\text{change in } x} = \frac{3}{2}$

y-intercept: $b = {}^{-}3$. The y-intercept is the point $(0,{}^{-}3)$.
Starting at $(0,{}^{-}3)$, the slope tells you to go up 3 and right 2 for (2,0).
Use a **solid** line for the border and shade **above** the line because the symbol is $\geq$.

* * * * * * * * * * * *

$y \geq 5 - \frac{5}{2}x$

Use the commutative property. $y \geq {}^{-}\frac{5}{2}x + 5$

slope: $m = \frac{\text{change in } y}{\text{change in } x} = \frac{-5}{2}$

y-intercept: $b = 5$. The y-intercept is the point (0,5).
Starting at (0,5), the slope tells you to go down 5 and right 2 for (2,0).
Use a **solid** line for the border and shade **above** the line because the symbol is $\geq$.

The solution for the system of inequalities is the double-shaded area on the graph.

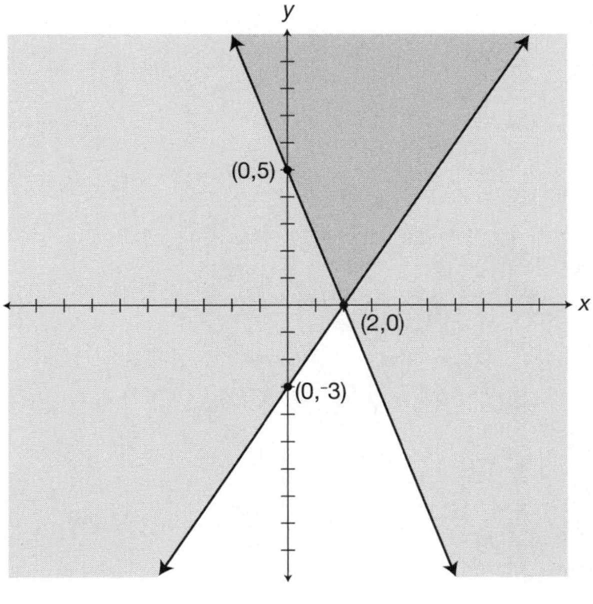

267. Transform equations into slope/y-intercept form. $6y < 5x - 30$

Divide both sides by 6. $y < \frac{5}{6}x - 5$

<u>slope:</u> $m = \frac{\text{change in } y}{\text{change in } x} = \frac{5}{6}$

<u>y-intercept:</u> $b = {}^-5$. The y-intercept is the point $(0,{}^-5)$.
Starting at $(0,{}^-5)$, the slope tells you to go up 5 and right 6 for (6,0).
Use a **dotted** line for the border and shade **below** it because the symbol is <.

* * * * * * * * * * * *

$2y < {}^-x + 4$

Divide both sides by 2. $y < \frac{^-1}{2}x + 2$

<u>slope:</u> $m = \frac{\text{change in } y}{\text{change in } x} = \frac{^-1}{2}$

<u>y-intercept:</u> $b = 2$. The y-intercept is the point (0,2).
Starting at (0,2), the slope tells you to go down 1 space and right 2 for (2,1).
Use a **dotted** line for the border and shade **below** it because the symbol is <.

<u>The solution for the system of inequalities is the double-shaded area on the graph.</u>

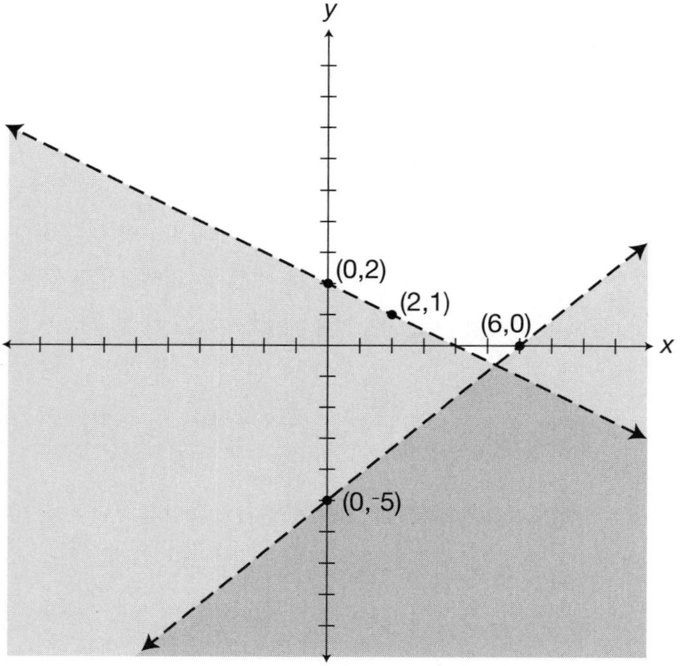

268. Transform equations into slope/y-intercept form. $y - x \geq 6$

Add x to both sides. $y - x + x \geq x + 6$

Simplify. $y \geq x + 6$

slope: $m = \frac{\text{change in } y}{\text{change in } x} = 1 = \frac{1}{1}$

y-intercept: $b = 6$. The y-intercept is the point (0,6).
Starting at (0,6), the slope tells you to go up 1 and right 1 for (1,7).
Use a **solid** line for the border and shade **above** it because the symbol is ≥.

* * * * * * * * * * * *

Use the distributive property of multiplication. $11y \geq {}^{-}2(x + 11)$

Simplify. $11y \geq {}^{-}2x - 22$

Divide both sides by 11. $y \geq \frac{^{-}2}{11}x - 2$

slope: $m = \frac{\text{change in } y}{\text{change in } x} = \frac{^{-}2}{11}$

y-intercept: $b = {}^{-}2$. The y-intercept is the point (0,⁻2).
Starting at (0,⁻2), the slope tells you to go down 2 and right 11 for (11,⁻4).
Use a **solid** line for the border and shade **above** the line because the symbol is ≥.

The solution for the system of inequalities is the double-shaded area on the graph.

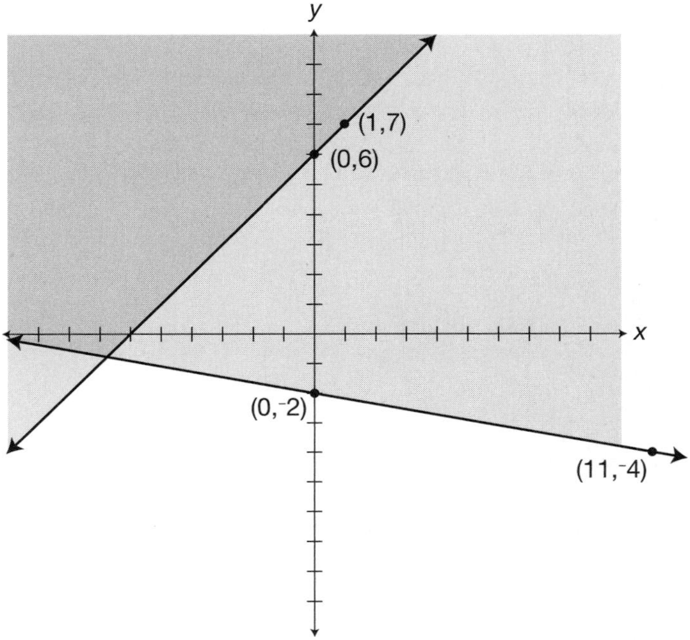

269. Transform equations into slope/y-intercept form. $5y \leq 8 = (x + 5)$

Use the distributive property of multiplication. $5y \leq 8x + 40$

Divide both sides by 5. $y \leq \frac{8}{5}x + 8$

<u>slope:</u> $m = \frac{\text{change in } y}{\text{change in } x} = \frac{8}{5} = \frac{^-8}{^-5}$

{Use the negatives to keep the coordinates near the origin.}

<u>y-intercept:</u> $b = 8$. The y-intercept is the point (0,8).
Starting at (0,8), the slope tells you to go down 8 and left 5 for ($^-5$,0).
Use a **solid** line for the border and shade **below** it because the symbol is ≤.

* * * * * * * * * * * *

$5y < 12(5 - x)$

Use the distributive property of multiplication. $5y < 60 - 12x$

Use the commutative property of addition. $5y < {^-12}x + 60$

Divide both sides by 5. $y < \frac{^-12}{5}x + 12$

<u>slope:</u> $m = \frac{\text{change in } y}{\text{change in } x} = \frac{^-12}{5}$

<u>y-intercept:</u> $b = 12$. The y-intercept is the point (0,12).
Starting at (0,12), the slope tells you to go down 12 and right 5 for (5,0).
Use a **dotted** line for the border and shade **below** the line because the symbol is <.

<u>The solution for the system of inequalities is the double-shaded area on the graph.</u>

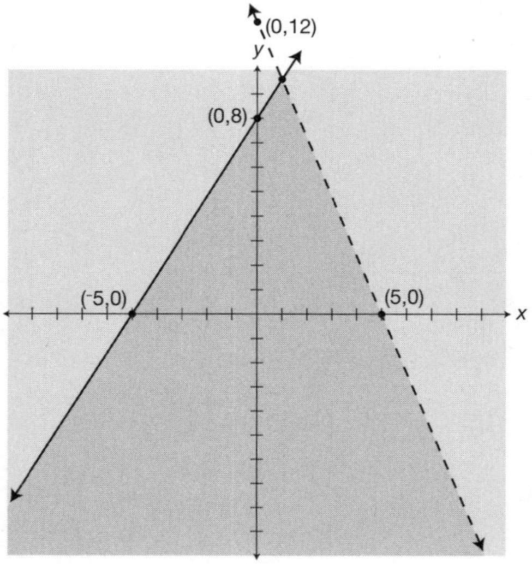

270. Transform equations into slope/y-intercept form. $2(x + 5y) > 5(x + 6)$

Use the distributive property of multiplication. $2x + 10y > 5x + 30$

Subtract $2x$ from both sides. $2x - 2x + 10y > 5x - 2x + 30$

Simplify the inequality. $10y > 3x + 30$

Divide both sides by 10. $y > \frac{3}{10}x + 3$

slope: $m = \frac{\text{change in } y}{\text{change in } x} = \frac{3}{10}$

y-intercept: $b = 3$. The y-intercept is the point (0,3).

Starting at (0,3), the slope tells you to go up 3 and right 10 for (10,6).

Use a **dotted** line for the border and shade **above** it because the symbol is >.

* * * * * * * * * * * *

$4x + y < 4x + 5$

Subtract $4x$ from both sides. $4x - 4x + y < 4x - 4x + 5$

Simplify. $y < 0x + 5$

With a slope of zero, the line is parallel to the x-axis.

The y-intercept is (0,5).

Use a **dotted** line for the border and shade **below** it because the symbol is <.

The solution for the system of inequalities is the double-shaded area on the graph.

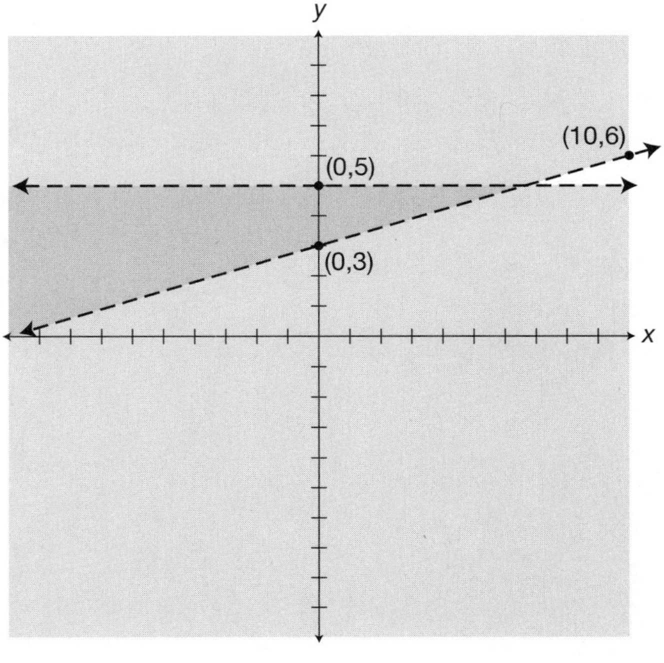

271. Transform equations into slope/y-intercept form. $3y \leq {}^{-}2(x + 3)$

Use the distributive property of multiplication. $3y \leq {}^{-}2x - 6$

Divide both sides by 3. $y \leq \frac{-2}{3}x - 2$

slope: $m = \frac{\text{change in } y}{\text{change in } x} = \frac{-2}{3}$

y-intercept: $b = {}^{-}2$. The y-intercept is the point $(0,{}^{-}2)$.
Starting at $(0,{}^{-}2)$, the slope tells you to go down 2 and right 3 for $(3,{}^{-}4)$.
Use a **solid** line for the border and shade **below** it because the symbol is $\leq$.

* * * * * * * * * * * *

Use the distributive property of multiplication. $3y \geq 12 - 2x$

Use the commutative property. $3y \geq {}^{-}2x + 12$

Divide both sides by 3. $y \geq \frac{-2}{3}x + 4$

slope: $m = \frac{\text{change in } y}{\text{change in } x} = \frac{-2}{3}$

y-intercept: $b = 4$. The y-intercept is the point $(0,4)$.
Starting at $(0,4)$, the slope tells you to go down 2 and right 3 for $(3,2)$.
Use a **solid** line for the border and shade **above** the line because the symbol is $\geq$.
{Slopes that are the same will result in parallel lines.}

The solution for the system of inequalities is the empty set because the areas don't overlap.

272. Transform equations into slope/y-intercept form. $9(y - 4) < 4x$

Use the distributive property of multiplication. $9y - 36 < 4x$

Add 36 to both sides. $9y - 36 + 36 < 4x + 36$

Divide both sides by 9. $y < \frac{4}{9}x + 4$

slope: $m = \frac{\text{change in } y}{\text{change in } x} = \frac{4}{9}$

y-intercept: $b = 4$. The y-intercept is the point (0,4).
Starting at (0,4), the slope tells you to go up 4 and right 9 for (9,8).
Use a **dotted** line for the border and shade **below** it because the symbol is <.

* * * * * * * * * * * *

$^-9y < 2(x + 9)$

Use the distributive property of multiplication. $^-9y < 2x + 18$

Divide both sides by $^-9$. Change the direction of the symbol when dividing by a negative. $y > \frac{^-2}{9}x - 2$

slope: $m = \frac{\text{change in } y}{\text{change in } x} = \frac{^-2}{9}$

y-intercept: $b = {^-2}$. The y-intercept is the point $(0,{^-2})$.
Starting at $(0,{^-2})$, the tells you to go down 2 and right 9 for $(9,{^-4})$.
Use a **dotted** line for the border and shade **above** the line because the symbol is >.

The solution for the system of inequalities is the double-shaded area on the graph.

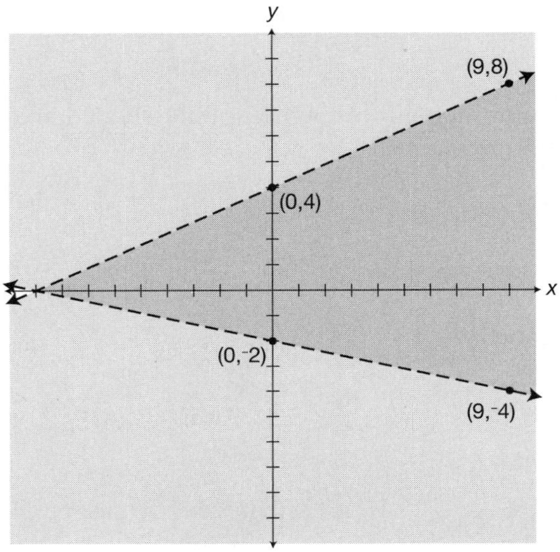

273. Transform equations into slope/y-intercept form. $7(y - 5) < {}^{-}5x$

Use the distributive property of multiplication. $7y - 35 < {}^{-}5x$

Add 35 to both sides. $7y - 35 + 35 < {}^{-}5x + 35$

Simplify. $7y < {}^{-}5x + 35$

Divide both sides by 7. $y < \frac{-5}{7}x + 5$

slope: $m = \frac{\text{change in } y}{\text{change in } x} = \frac{-5}{7}$

y-intercept: $b = 5$. The y-intercept is the point (0,5).
Starting at (0,5), the slope tells you to go down 5 and right 7 for (7,0).
Use a **dotted** line for the border and shade **below** it because the symbol is <.

* * * * * * * * * * * *

$^{-}3 < \frac{1}{4}(2x - 3y)$

Multiply both sides of the inequality by 4. $4({}^{-}3) < 4(\frac{1}{4})(2x - 3y)$

Simplify the inequality. $^{-}12 < 1(2x - 3y)$

$^{-}12 < 2x - 3y$

Add $3y$ to both sides. $^{-}12 + 3y < 2x - 3y + 3y$

Simplify the inequality. $^{-}12 + 3y < 2x$

Add 12 to both sides. $^{-}12 + 12 + 3y < 2x + 12$

Simplify the inequality. $3y < 2x + 12$

Divide both sides by 3. $y < \frac{2}{3}x + 4$

slope: $m = \frac{\text{change in } y}{\text{change in } x} = \frac{2}{3}$

y-intercept: $b = 4$. The y-intercept is the point (0,4).
Starting at (0,4), the slope tells you to go up 2 and right 3 for (3,6).
Use a **dotted** line for the border and shade **below** the line because the symbol is <.

The solution for the system of inequalities is the double-shaded area on the graph (next page).

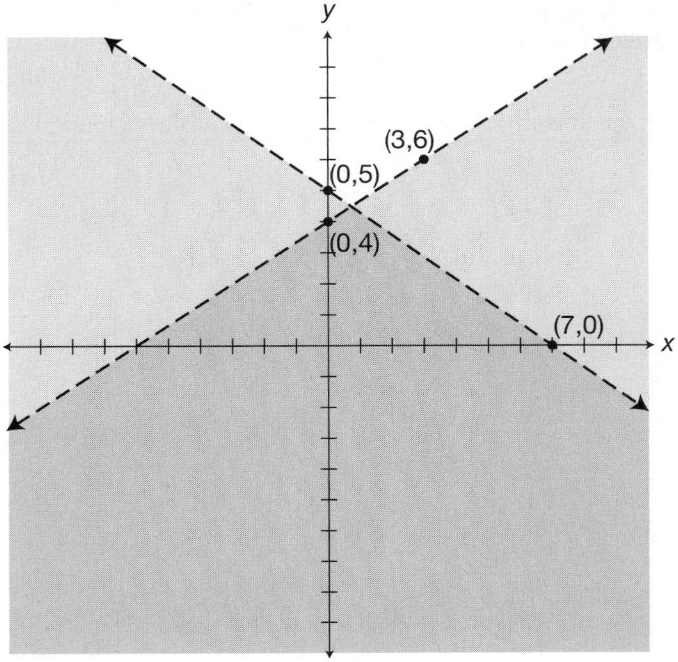

Graph for question 273

274. Transform equations into slope/y-intercept form. $y > \frac{7}{4}(4 - x)$

Use the distributive property of multiplication. $y > 7 - \frac{7}{4}x$

Use the commutative property. $y > \frac{^-7}{4}x + 7$

slope: $m = \frac{\text{change in } y}{\text{change in } x} = \frac{^-7}{4}$

y-intercept: $b = 7$. The y-intercept is the point (0,7).
Starting at (0,7), the slope tells you to go down 7 and right 4 for (4,0).
Use a **dotted** line for the border and shade **above** it because the symbol is >.

* * * * * * * * * * * *

$3(y + 5) > 7x$

Use the distributive property of multiplication. $3y + 15 > 7x$

Subtract 15 from both sides. $3y + 15 - 15 > 7x - 15$

Simplify the inequality. $3y > 7x - 15$

Divide both sides by 3. $y > \frac{7}{3}x - 5$

slope: $m = \frac{\text{change in } y}{\text{change in } x} = \frac{7}{3}$

y-intercept: $b = {^-5}$. The y-intercept is the point $(0,{^-5})$.

Starting at $(0,^-5)$, the slope tells you to go up 7 and right 3 for (3,2).
Use a **dotted** line for the border and shade **above** the line because the symbol is >.

The solution for the system of inequalities is the double-shaded area on the graph.

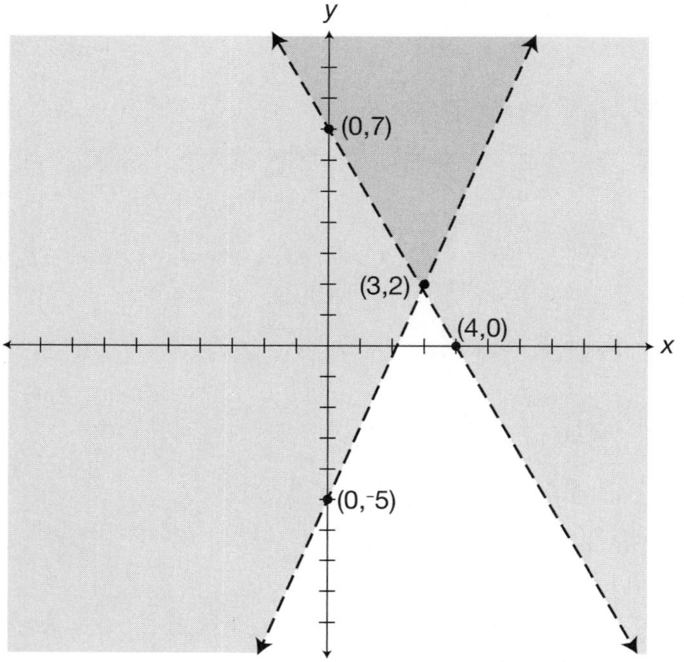

Graph for question 274

275. Transform equations into slope/y-intercept form. $5x - 2(y + 10) \leq 0$
Use the distributive property of multiplication. $5x - 2y - 20 \leq 0$
Subtract $5x$ from both sides. $5x - 5x - 2y - 20 \leq {}^-5x$
Simplify. ${}^-2y - 20 \leq {}^-5x$
Add 20 to both sides. ${}^-2y - 20 + 20 \leq {}^-5x + 20$
Simplify the inequality. ${}^-2y \leq {}^-5x + 20$
Divide both sides by $^-2$. Change the direction of the symbol when dividing by a negative. $y \geq \frac{5}{2}x - 10$

slope: $m = \frac{\text{change in } y}{\text{change in } x} = \frac{5}{2}$

y-intercept: $b = {}^-10$. The y-intercept is the point $(0,^-10)$.
Starting at $(0,^-10)$, the slope tells you to go up 5 and right 2 for $(2,^-5)$.
Use a **solid** line for the border and shade **above** it because the symbol is ≥.

* * * * * * * * * * * *

	$2x + y \leq {}^-3$
Subtract $2x$ from both sides.	$2x - 2x + y \leq {}^-2x - 3$
Simplify the inequality.	$y \leq {}^-2x - 3$

slope: $m = \frac{\text{change in } y}{\text{change in } x} = \frac{^-2}{1}$

y-intercept: $b = {}^-3$. The *y*-intercept is the point $(0,^-3)$.
Starting at $(0,^-3)$, the slope tells you to go down 2 and right 1 for $(1,^-5)$.
Use a **solid** line for the border and shade **below** the line because the symbol is $\leq$.

The solution for the system of inequalities is the double-shaded area on the graph.

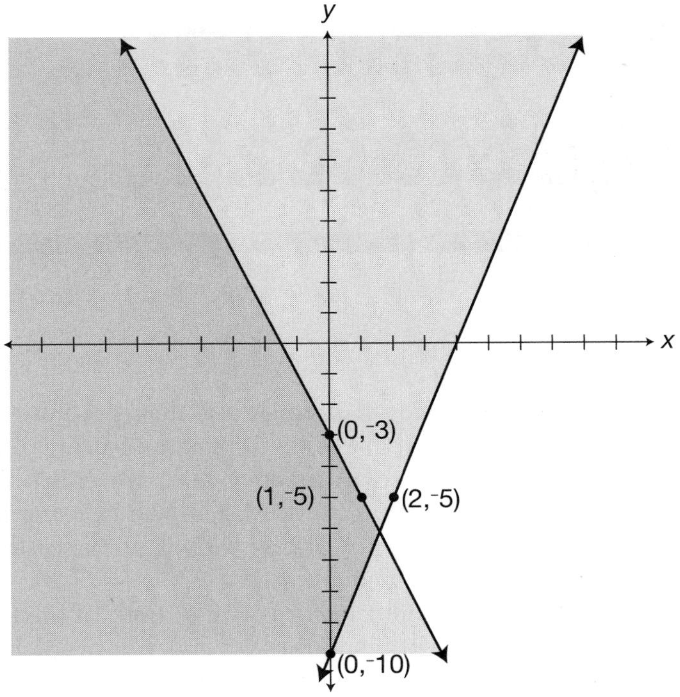

Graph for question 275

12

Solving Systems of Equations Algebraically

There is a faster way to solve systems of equations than graphing to locate the solution points. You can solve systems of equations using algebraic methods. The two methods you will practice here are called the **elimination method** and the **substitution method**. You will be using the skills you have practiced in the chapters on working with algebraic expressions, combining like terms, and solving equations.

In the **elimination method,** you will transform one or both of the two equations in the system so that when you add the two equations together, one of the variables will be eliminated. Then you solve the resulting equation for the remaining variable. When you find a numerical value for the remaining variable, you finally substitute the found value into one of the two equations to determine the value of the remaining variable.

In the **substitution method**, you will transform one of the equations so that one variable is expressed in terms of the other. Then you will eliminate the variable by substituting this expression into the other equation and solving for the remaining variable. When you find a numerical value for one variable, use it in one of the two equations to determine the value of the remaining variable.

One method is not better than the other. But you may find that you will begin to see which equations, because of their structure, lend themselves to one method over the other. Practice will help you decide.

Tips for Solving Systems of Equations Algebraically

When using the elimination method, first make a plan to determine which variable you will eliminate from the system. Then transform the equation or equations so that you will get the result you want.

Express your solution as a coordinate point of the form (x,y).

Use the elimination method to solve the following systems of equations.

276. $x + y = 4$
$2x - y = {}^-1$

277. $3x + 4y = 17$
${}^-x + 2y = 1$

278. $7x + 3y = 11$
$2x + y = 3$

279. $4x - 0.5y = {}^-16$
${}^-6x + y = 8$

280. $3(x + y) = 18$
$5x + y = {}^-2$

281. ${}^-2(x - \frac{1}{2}y) = 3$
$4(2y - 3x) = 8$

282. $6y + 3x = 30$
$2y + 6x = 0$

283. $5x + 8y = 25$
$3x - 15 = y$

284. $3x = 5 - 7y$
$2y = x - 6$

285. $3x + y = 20$
$\frac{x}{3} + 10 = y$

286. $3x - 5y = {}^-21$
$2(2y - x) = 16$

287. $3({}^-2x + 3y) = {}^-1$
$4(2y - x) = 1$

288. $\frac{1}{4}x + y = 12$
$2x - \frac{1}{3}y = 21$

Use the substitution method to solve the following systems of equations.

289. $y = 5x$
$4x + 5y = 87$

290. $x + y = 3$
$3x + 101 = 7y$

291. $5x + y = 3.6$
$y + 21x = 8.4$

292. $8y - x = 0$
$10y + x = 36$

293. $\frac{x}{3} = y + 2$
$2x - 4y = 32$

294. $x + 4y = 0$
$^{-}4y + x = 24$

295. $5x + y = 20$
$3x = \frac{1}{2}y + 1$

296. $2x + y = 2 - 5y$
$x - y = 5$

297. $\frac{2x}{10} + \frac{y}{5} = 1$
$3x + 2y = 12$

298. $4y + 31 = 3x$
$y + 10 = 3x$

299. $y = 2x + 5$
$1 - 3y = 2x$

300. $2(2 - x) = 3y - 2$
$3x + 9 = 4(5 - y)$

Answers

Numerical expressions in parentheses like this [ ] are operations performed on only part of the original expression. The operations performed within these symbols are intended to show how to evaluate the various terms that make up the entire expression.

Expressions with parentheses that look like this ( ) contain either numerical substitutions or expressions that are part of a numerical expression. Once a single number appears within these parentheses, the parentheses are no longer needed and need not be used the next time the entire expression is written.

When two pair of parentheses appear side by side like this ( )( ), it means that the expressions within are to be multiplied.

Sometimes parentheses appear within other parentheses in numerical or algebraic expressions. Regardless of what symbol is used, ( ), { }, or [ ], perform operations in the innermost parentheses first and work outward.

The underlined ordered pair is the solution.

276. Add the equations. $x + y = 4$
$2x - y = {}^-1$
$3x + 0 = 3$

Additive identity. $3x = 3$

Now solve for x.

Divide both sides by 3. $\frac{3x}{3} = \frac{3}{3}$

Simplify terms. $x = 1$

Substitute the value of x into one of the original equations in the system and solve for y. $(1) + y = 4$

Subtract 1 from both sides. $1 - 1 + y = 4 - 1$

Simplify. $y = 3$

The solution for the system of equations is (1,3).

277. We could add the equations together if we had a ^-3x in the second equation.

Multiply the second equation by 3. $3({}^-x + 2y = 1)$

Simplify. $^-3x + 6y = 3$

Add the first and transformed second equations. $3x + 4y = 17$
$0 + 10y = 20$

Identity element of addition. $10y = 20$

Divide both sides by 10. $y = 2$

Substitute the value of y into one of the original equations in the system and solve for x. $^-x + 2(2) = 1$

Simplify. $^-x + 4 = 1$

Subtract 4 from both sides. $^-x + 4 - 4 = 1 - 4$

Simplify. $^-x = {}^-3$

Multiply both sides by $^-1$. $x = 3$

The solution for the system of equations is (3,2).

278. Transform the second equation so you can add it to the first and eliminate y.

Multiply the equation by $^-3$.	$^-3(2x + y = 3)$
Simplify.	$^-6x - 3y = {}^-9$
Add the first equation to the transformed second.	
	$^-6x - 3y = {}^-9$
	$\underline{7x + 3y = 11}$
	$x + 0 = 2$
	$x = 2$
Substitute the value of x into one of the original equations in the system and solve for y.	$2(2) + y = 3$
	$4 + y = 3$
Subtract 4 from both sides.	$4 - 4 + y = 3 - 4$
Simplify.	$y = {}^-1$

The solution for the system of equations is $(2,{}^-1)$.

279. If you multiply the first equation by 2 and add the equations together, you can eliminate the y.

	$2(4x - 0.5y) = 2(^-16)$
Use the distributive property of multiplication.	$2(4x) - 2(0.5y) = 2(^-16)$
Simplify terms.	$8x - y = {}^-32$
Add this equation to the given second equation.	$8x - y = {}^-32$
	$\underline{^-6x + y = 8}$
	$2x + 0 = {}^-24$
Additive identity.	$2x = {}^-24$
Divide both sides by 2.	$x = {}^-12$
Substitute the value of into one of the equations in the system and solve for y.	$^-6(^-12) + y = 8$
Simplify.	$72 + y = 8$
Subtract 72 from both sides.	$72 - 72 + y = 8 - 72$
Simplify.	$y = {}^-64$

The solution for the system of equations is $(^-12, {}^-64)$.

280. See what the first equation looks like after distributing the multiplication on the left.

Use the distributive property of multiplication.	$3x + 3y = 18$
Multiply the second equation by $^-3$ and add it to the first to eliminate y.	$^-3(5x + y) = {}^-3(^-2)$
Use the distributive property of multiplication.	$^-3(5x) - 3(y) = {}^-3(^-2)$
Simplify terms.	$^-15x - 3y = 6$
Add the transformed first equation.	$\underline{3x + 3y = 18}$
	$^-12x + 0 = 24$
Additive identity.	$^-12x = 24$

Divide both sides by $^{-}12$. $x = {}^{-}2$

Substitute the value of x into one of the original equations in the system and solve for y.

$5({}^{-}2) + y = {}^{-}2$

$^{-}10 + y = {}^{-}2$

Add 10 to both sides. $10 - 10 + y = 10 - 2$

Simplify. $y = 8$

The solution for the system of equations is $({}^{-}2,8)$.

281. See what the equations look like after applying the distributive law for multiplication on the left side of each of them.

$$^{-}2(x) - 2\left({}^{-}\tfrac{1}{2}y\right) = 3$$

$$^{-}2x + y = 3$$

$$4(2y) - 4(3x) = 8$$

$$8y - 12x = 8$$

If you multiply the first of the transformed equations by $^{-}8$ and add it to the second of the transformed equations, you can eliminate y. $^{-}8({}^{-}2x + y) = {}^{-}8(3)$

Use the distributive property of multiplication. $^{-}8({}^{-}2x) - 8(y) = {}^{-}8(3)$

Simplify terms. $16x - 8y = {}^{-}24$

Use the commutative law of addition to switch the order of terms on the left side. $^{-}8y + 16x = {}^{-}24$

Add this equation to the given second equation.

$$\begin{array}{r} 8y - 12x = 8 \\ \underline{{}^{-}8y + 16x = {}^{-}24} \\ 0y + 4x = {}^{-}16 \end{array}$$

Additive identity. $4x = {}^{-}16$

Divide both sides by 4. $x = {}^{-}4$

Substitute the value of x into one of the equations in the system and solve for y. $4(2y - 3({}^{-}4)) = 8$

Simplify.

$$4(2y + 12) = 8$$

$$4(2y) + 4(12) = 8$$

$$8y + 48 = 8$$

Subtract 48 from both sides. $8y + 48 - 48 = 8 - 48$

Simplify. $8y = {}^{-}40$

Divide both sides by 8. $y = {}^{-}5$

The solution for the system of equations is $({}^{-}4,{}^{-}5)$.

282. Multiply the second equation by $^-3$ and add it to the first equation to eliminate y.

$^-3(2y + 6x = 0)$
$^-6y - 18x = 0$
$\underline{6y + 3x = 30}$
$0 - 15x = 30$

Additive identity. $^-15x = 30$

Divide both sides of the equation by $^-15$. $x = {}^-2$

Substitute the value of x into one of the original equations in the system and solve for y.

$2y + 6(^-2) = 0$
$2y - 12 = 0$

Add 12 to both sides. $2y - 12 + 12 = 0 + 12$

Simplify. $2y = 12$

Divide both sides by 2. $y = 6$

The solution for the system of equations is $(^-2,6)$.

283. Transform the second equation into a similar format to the first equation, then line up like terms.

Add 15 to both sides. $3x - 15 + 15 = y + 15$

Simplify. $3x = y + 15$

Subtract y from both sides. $3x - y = y - y + 15$

Simplify. $3x - y = 15$

Multiply this version of the second equation by 8 and add the first equation. $8(3x - y = 15)$

Use the distributive property of multiplication. $24x - 8y = 120$

Add the first equation to the second. $\underline{5x + 8y = 25}$

$29x + 0 = 145$

Additive identity. $29x = 145$

Divide both sides by 29. $x = 5$

Substitute the value of x into one of the original equations in the system and solve for y. $3(5) - 15 = y$

Simplify. $15 - 15 = y$

$0 = y$

The solution for the system of equations is $(5,0)$.

284. Transform the first equation into familiar form ($ax + by = c$). $3x = 5 - 7y$

Add $7y$ to both sides. $3x + 7y = 5 - 7y + 7y$

Simplify. $3x + 7y = 5$

Transform the second equation into familiar form ($ax + by = c$). $2y = x - 6$

Subtract x from both sides. $^-x + 2y = x - x - 6$

Simplify. $^-x + 2y = {^-6}$

Multiply this equation by 3. $3(^-x + 2y = {^-6})$

$3(^-x) + 3(2y) = 3(^-6)$

Simplify terms. $^-3x + 6y = {^-18}$

Add this equation to the transformed first equation. $3x + 7y = 5$

$0 + 13y = {^-13}$

Additive identity. $13y = {^-13}$

Divide both sides of the equation by 13. $y = {^-1}$

Substitute the value of y into one of the original equations in the system and solve for x. $3x = 5 - 7(^-1)$

$3x = 5 + 7$

$3x = 12$

Divide both sides by 3. $x = 4$

The solution for the system of equations is $(4,{^-1})$.

285. Transform the second equation into familiar form ($ax + by = c$).

Multiply the equation by 3. $3(\frac{x}{3} + 10 = y)$

Use the distributive property. $3(\frac{x}{3}) + 3(10) = 3(y)$

Simplify terms. $x + 30 = 3y$

Subtract 30 from each side. $x + 30 - 30 = 3y - 30$

Simplify. $x = 3y - 30$

Subtract $3y$ from both sides. $x - 3y = 3y - 3y - 30$

Simplify. $x - 3y = {^-30}$

Multiply the first equation by 3 and add to the transformed second equation to eliminate y. $3(3x + y = 20)$

Use the distributive property. $3(3x) + 3(y) = 3(20)$

Simplify terms. $9x + 3y = 60$

Add the transformed second equation to the first. $x - 3y = {^-30}$

$10x + 0 = 30$

Additive identity. $10x = 30$

Divide both sides of the equation by 10. $x = 3$

Substitute the value of x into one of the original equations in the system and solve for y. $\frac{(3)}{3} + 10 = y$

Simplify terms. $1 + 10 = y$

$11 = y$

The solution for the system of equations is $(3,11)$.

286. Transform the second equation into a similar form to the first equation.

Use the distributive property of multiplication.	$2(2y) - 2(x) = 16$
Simplify terms.	$4y - 2x = 16$
Commutative property of addition.	$^-2x + 4y = 16$
Multiply the first equation by 2 and the second equation by 3, and add the transformed equations to eliminate the variable x.	$2(3x - 5y = {}^-21)$
Distributive property. Simplify terms.	$6x - 10y = {}^-42$
	$3({}^-2x + 4y = 16)$
Distributive property. Simplify terms.	$^-6x + 12y = 48$
Add the first equation to the second.	$6x - 10y = {}^-42$
	$2y = 6$
Divide both sides by 2.	$y = 3$
Substitute the value of y into one of the original equations in the system and solve for x.	$3x - 5(3) = {}^-21$
Simplify terms and add 15 to each side.	$3x - 15 + 15 = {}^-21 + 15$
Combine like terms on each side.	$3x = {}^-6$
Divide both sides by 3.	$x = {}^-2$

The solution for the system of equations is $({}^-2,3)$.

287.

See what the equations look like after applying the distributive law for multiplication on the left side of each of them.	$3({}^-2x) + 3(3y) = {}^-1$
	$^-6x + 9y = {}^-1$
	$4(2y) - 4(x) = 1$
	$8y - 4x = 1$
If you multiply the first of the transformed equations by 2 and the second by $^-3$ and then add, you can eliminate x.	$2({}^-6x + 9y) = 2({}^-1)$
	$^-3(8y - 4x) = {}^-3(1)$
Use the distributive property of multiplication.	$2({}^-6x) + 2(9y) = 2({}^-1)$
	$^-3(8y) - 3({}^-4x) = {}^-3(1)$
Simplify terms.	$^-12x + 18y = {}^-2$
	$^-24y + 12x = {}^-3$
Use the commutative law of addition to switch the order of terms on the left side of the first equation.	$18y - 12x = {}^-2$
	$^-24y + 12x = {}^-3$
Add this equation to the given second equation.	$^-6y + 0 = {}^-5$
Additive identity.	$^-6y = {}^-5$
Divide both sides by $^-6$.	$y = \frac{5}{6}$

Substitute the value of y into one of the equations in the system and solve for x.	$3(^-2x + 3(\frac{5}{6})) = {}^-1$
Simplify.	$3(^-2x) + 3(3(\frac{5}{6})) = {}^-1$
	$^-6x + \frac{15}{2} = {}^-1$
Subtract $\frac{15}{2}$ from both sides.	$^-6x + \frac{15}{2} - \frac{15}{2} = {}^-1 - \frac{15}{2}$
Simplify.	$^-6x = {}^-\frac{17}{2}$
Divide both sides by $^-6$.	$x = \frac{17}{12}$

The solution for the system of equations is $(\frac{17}{2}, \frac{5}{6})$.

288. Transform the second equation by multiplying it by 3.

Then, add the equations together to eliminate y.	$3(2x - \frac{1}{3}y = 21)$
Use the distributive property of multiplication.	$3(2x) - 3(\frac{1}{3}y) = 3(21)$
Simplify terms.	$6x - y = 63$
Add the first equation to the second.	$\frac{1}{4}x + y = 12$
	$6\frac{1}{4}x + 0 = 75$
Additive identity.	$6\frac{1}{4}x = 75$
Divide both sides by $6\frac{1}{4}$.	$\frac{6\frac{1}{4}x}{6\frac{1}{4}} = \frac{75}{6\frac{1}{4}}$
Simplify.	$x = 12$
Substitute the value of x into one of the original equations in the system and solve for y.	$\frac{1}{4}(12) + y = 12$
Simplify the first term and subtract 3 from both sides.	$3 - 3 + y = 12 - 3$
Simplify.	$y = 9$

The solution for the system of equations is (12,9).

289. The first equation tells you that $y = 5x$. Substitute $5x$ for y in the second equation and then solve for x. $4x + 5(5x) = 87$

Simplify term and add like terms.	$4x + 25x = 87$
	$29x = 87$
Divide both sides by 29.	$x = \frac{87}{29} = 3$
Substitute 3 for x in one of the original equations.	$y = 5 \cdot (3) = 15$

The solution for the system of equations is (3,15).

290. Transform the first equation so that the value of x is expressed in terms of y.
Subtract y from both sides of the equation. $x + y - y = 3 - y$
Simplify. $x = 3 - y$
Substitute $3 - y$ for x in the second equation and solve for y. $3(3 - y) + 101 = 7y$
Use the distributive property of multiplication. $9 - 3y + 101 = 7y$
Use the commutative property of addition. $9 + 101 - 3y = 7y$
Add like terms. Add $3y$ to both sides. $110 - 3y + 3y = 7y + 3y$
Combine like terms. $110 = 10y$
Divide both sides by 10. $11 = y$
Substitute the value of y into one of the original equations in the system and solve for x. $x + (11) = 3$
Subtract 11 from both sides. $x + 11 - 11 = 3 - 11$
Combine like terms on each side. $x = {}^{-}8$
The solution for the system of equations is $({}^{-}8,11)$.

291. Transform the first equation so that y is expressed in terms of x. $5x + y = 3.6$
Subtract $5x$ from both sides of the equation. $5x - 5x + y = 3.6 - 5x$
Combine like terms on each side. $y = 3.6 - 5x$
Substitute the value of y into the second equation. $(3.6 - 5x) + 21x = 8.4$
Combine like terms. $3.6 + 16x = 8.4$
Subtract 3.6 from both sides. $3.6 - 3.6 + 16x = 8.4 - 3.6$
Combine like terms on each side. $16x = 4.8$
Divide both sides by 16. $x = 0.3$
Substitute the value of x into one of the original equations in the system and solve for y. $5(0.3) + y = 3.6$
Simplify terms. $1.5 + y = 3.6$
Subtract 1.5 from both sides. $1.5 - 1.5 + y = 3.6 - 1.5$
Combine like terms on each side. $y = 2.1$
The solution for the system of equations is $(0.3,2.1)$.

292. Solve the first equation for x. $x = 8y$
Substitute this expression for x into the second equation. $10y + (8y) = 36$
Simplify. $18y = 36$
Divide both sides by 18. $y = 2$
Substitute the value of y into one of the equations in the system and solve for x. $8(2) - x = 0$
Simplify. $16 - x = 0$
Add x to both sides of the equation. $16 - x + x = 0$
Simplify. $16 = x$
The solution for the system of equations is $(16,2)$.

293. Transform the first equation so that the value of x is expressed in terms of y.

Multiply the equation by 3.	$3((\frac{x}{3}) = y + 2)$
Use the distributive property.	$3(\frac{x}{3}) = 3y + 6$
Simplify.	$x = 3y + 6$
Substitute the value of x into the second equation in the system and solve for y.	$2(3y + 6) - 4y = 32$
Use the distributive property of multiplication.	$6y + 12 - 4y = 32$
Use the commutative property of addition.	$6y - 4y + 12 = 32$
Combine like terms. Subtract 12 from both sides.	$2y + 12 - 12 = 32 - 12$
Combine like terms on each side.	$2y = 20$
Divide both sides by 2.	$y = 10$
Substitute the value of y into one of the original equations in the system and solve for x.	$2x - 4(10) = 32$
Simplify and add 40 to both sides.	$2x - 40 + 40 = 32 + 40$
Combine like terms.	$2x = 72$
Divide both sides by 2.	$x = 36$

The solution for the system of equations is (36,10).

294.

Solve the first equation for x.	$x = {}^-4y$
Substitute this expression for x into the second equation.	${}^-4y + ({}^-4y) = 24$
Simplify.	${}^-8y = 24$
Divide both sides by ${}^-8$.	$y = {}^-3$
Substitute the value of y into one of the equations in the system and solve for x.	$x + 4({}^-3) = 0$
Simplify.	$x - 12 = 0$
Add 12 to both sides of the equation.	$x - 12 + 12 = 0 + 12$
Simplify.	$x = 12$

The solution for the system of equations is (12,${}^-3$).

295.

Transform the first equation so that the value of y is expressed in terms of x. Subtract $5x$ from both sides of the equation.	$5x - 5x + y = 20 - 5x$
Combine like terms.	$y = 20 - 5x$
Substitute the value of y into the second equation in the system and solve for x.	$3x = \frac{1}{2}(20 - 5x) + 1$
Use the distributive property of multiplication.	$3x = 10 - \frac{5}{2}x + 1$
Combine like terms.	$3x = 11 - \frac{5}{2}x$
Add $\frac{5}{2}x$ to both sides.	$3x + \frac{5}{2}x = 11 + \frac{5}{2}x - \frac{5}{2}x$
Combine like terms.	$5\frac{1}{2}x = 11$
Divide both sides by $5\frac{1}{2}$.	$\frac{5\frac{1}{2}x}{5\frac{1}{2}} = \frac{11}{5\frac{1}{2}}$

Simplify terms. $x = 2$
Substitute the found value for x into one of the original equations and solve for y. $5(2) + y = 20$
Simplify. $10 + y = 20$
Subtract 10 from both sides. $y = 10$
The solution for the system of equations is (2,10).

296. Transform the second equation so that the value of x is expressed in terms of y. Add y to both sides of the equation. $x - y + y = 5 + y$
Combine like terms on each side. $x = 5 + y$
Substitute the value of x into the second equation in the system and solve for y. $2(5 + y) + y = 2 - 5y$
Use the distributive property of multiplication. $10 + 2y + y = 2 - 5y$
Add $5y$ to both sides of the equation. $10 + 2y + y + 5y = 2 - 5y + 5y$
Combine like terms on each side. $10 + 8y = 2$
Subtract 10 from both sides. $10 - 10 + 8y = 2 - 10$
Combine like terms on each side. $8y = {}^{-}8$
Divide both sides by 8. $y = {}^{-}1$
Substitute the found value for y into one of the original equations and solve for x. $x - ({}^{-}1) = 5$
Simplify. $x + 1 = 5$
$x = 4$
The solution for the system of equations is $(4,{}^{-}1)$.

297. Transform the first equation by eliminating the denominators.
Multiply both sides of the equation by 10. $10(\frac{2x}{10} + \frac{y}{5}) = 10(1)$
Use the distributive property of multiplication. $10(\frac{2x}{10}) + 10(\frac{y}{5}) = 10$
Simplify terms. $2x + 2y = 10$
Divide both sides by 2. $\frac{2x + 2y}{2} = \frac{10}{2}$
Simplify terms. $x + y = 5$
Now express x in terms of y. Subtract y from both sides of the equation. $x + y - y = 5 - y$
Simplify. $x = 5 - y$
Substitute the value of x into the second equation and solve for y. $3(5 - y) + 2y = 12$
Use the distributive property of multiplication. $3(5) - 3y + 2y = 12$
Combine like terms on each side. $15 - y = 12$
Add y to both sides. $15 - y + y = 12 + y$
Combine like terms. $15 = 12 + y$
Subtract 12 from both sides. $15 - 12 = 12 - 12 + y$
Simplify. $3 = y$
Substitute the value of y into one of the original equations in the system and solve for x. $3x + 2(3) = 12$

Simplify the term and subtract 6 from both sides. $3x + 6 - 6 = 12 - 6$
Combine like terms on each side. $3x = 6$
Divide both sides by 3. $x = 2$
The solution for the system of equations is (2,3).

298. Transform the second equation so that the value of y is expressed in terms of x.
Subtract 10 from both sides of the equation. $y + 10 - 10 = 3x - 10$
Combine like terms on each side. $y = 3x - 10$
Substitute the value of y into the first equation in the system and solve for x. $4(3x - 10) + 31 = 3x$
Use the distributive property of multiplication. $4(3x) - 4(10) + 31 = 3x$
Simplify terms. $12x - 40 + 31 = 3x$
Combine like terms on each side. $12x - 9 = 3x$
Add 9 to both sides of the equation. $12x - 9 + 9 = 3x + 9$
Combine like terms on each side. $12x = 3x + 9$
Subtract $3x$ from both sides. $12x - 3x = 3x - 3x + 9$
Combine like terms on each side. $9x = 9$
Divide both sides by 9. $x = 1$
Substitute the value of x into one of the original equation and solve for y. $y + 10 = 3(1)$
Subtract 10 from both sides. $y + 10 - 10 = 3 - 10$
Combine like terms on each side. $y = {}^-7$
The solution for the system of equations is $(1,{}^-7)$.

299. This time, each equation is solved for a different variable. As such, we can substitute the first into the second, or vice versa. Let's substitute the expression for y given by the first equation into the second. $1 - 3(2x + 5) = x$
Simplify. $1 - 3(2x) - 3(5) = x$
$1 - 6x - 15 = x$
${}^-6x - 14 = x$
Add 6 to both sides. $6x - 6x - 14 = x + 6x$
Simplify. ${}^-14 = 7x$
Divide both sides by 7. $x = {}^-2$
Substitute the value of into the first equation of the system and solve for y. $y = 2({}^-2) + 5$
Simplify. $y = 1$
The solution for the system of equations is $({}^-2,1)$.

300. Begin with the second equation and express x in terms of y.

Use the distributive property of multiplication.	$3x + 9 = 20 - 4y$
Subtract 9 from both sides.	$3x + 9 - 9 = 20 - 9 - 4y$
Combine like terms on each side.	$3x = 11 - 4y$
Divide both sides by 3.	$x = \frac{11 - 4y}{3}$
Substitute the value of x into the second equation and solve for y. First, use the distributive property to simplify the equation.	$4 - 2x = 3y - 2$
	$4 - 2(\frac{11 - 4y}{3}) = 3y - 2$
Multiply the numerator by the factor 2.	$4 - (\frac{22 - 8y}{3}) = 3y - 2$
Multiply both sides of the equation by 3 to eliminate the denominator.	$3(4 - (\frac{22 - 8y}{3})) = 3(3y - 2)$
Use the distributive property of multiplication.	$3(4) - 3(\frac{22 - 8y}{3}) = 3(3y) - 3(2)$
Simplify each term.	$12 - (22 - 8y) = 9y - 6$
Simplify the second term and the – sign.	$12 - 22 + 8y = 9y - 6$
Combine like terms.	$^-10 + 8y = 9y - 6$
Add 6 to both sides.	$6 - 10 + 8y = 9y + 6 - 6$
Combine like terms on each side.	$^-4 + 8y = 9y$
Subtract $8y$ from both sides.	$^-4 + 8y - 8y = 9y - 8y$
Combine like terms on each side.	$^-4 = y$
Substitute the value of y into one of the original equations in the system and solve for x.	$2(2 - x) = 3(^-4) - 2$
Distributive property of multiplication.	$4 - 2x = {}^-12 - 2$
Subtract 4 from both sides.	$4 - 4 - 2x = {}^-12 - 2 - 4$
Simplify.	$^-2x = {}^-18$
Divide both sides by $^-2$.	$x = 9$

The solution for the system of equations is $(9,{}^-4)$.

13 Working with Exponents

In this chapter, you will practice adding, subtracting, multiplying, and dividing expressions that contain variables with exponents. You will follow all the rules you have learned about operating with variables, but in this chapter, the variables now have exponents.

Tips for Working with Exponents

Add and subtract like terms:

$$3n + 5n = 8n, \text{ or } 5x^2y - 3x^2y = 2x^2y.$$

When multiplying variables with exponents, if the variables are the same, add the exponents and write the base only once:

$$(a^4)(a^3) = a^{(4+3)} = a^7$$
$$(x^2y^3)(axy^5) = ax^{(2+1)}y^{(3+5)} = ax^3y^8$$

When dividing variables with exponents, if the variables are the same, you subtract the exponents:

$$\frac{n^5}{n^2} = \frac{n \cdot n \cdot n \cdot n \cdot n}{n \cdot n} = n^{5-2} = n^3$$

If the exponent of a similar term in the denominator is larger than the one in the numerator, the exponent will have a negative sign:

$$\frac{2x^3}{x^4} = 2x^{-1}$$

$$\frac{n^5}{n^8} = n^{5-8} = n^{-3}$$

A negative power in the numerator becomes positive when the variable is moved into the denominator.

$$2x^{-1} = 2(\frac{1}{x^1}) = \frac{2}{x}$$

$$n^{-3} = \frac{1}{n^3}$$

When the result of a division leaves an exponent of zero, the term raised to the power of zero equals 1:

$$z^0 = 1$$

$$3\frac{r^2}{r^2} = 3r^0 = 3(1) = 3$$

When a variable with an exponent is raised to a power, you multiply the exponent to form the new term:

$$(b^2)^3 = b^2 \cdot b^2 \cdot b^2 = b^{(2+2+2)} = b^6$$

$$(2x^2y)^2 = 2x^2y \cdot 2x^2y = 2 \cdot 2 \cdot x^2 \cdot x^2 \cdot y \cdot y = 2^2x^{2+2}y^{1+1} = 4x^4y^2$$

Remember order of operations: PEMDAS. Generally, list terms in order from highest power to lowest power.

Simplify the following expressions:

301. $5x^2 + 8x^2$

302. $5ab^4 - ab^4$

303. $9mn^3 + 8mn + 2mn^3$

304. $5c^2 + 3c - 2c^2 + 4 - 7c$

305. $3x^2 + 4ax - 8a^2 + 7x^2 - 2ax + 7a^2$

306. $5xy \cdot 6xy + 7x^2y^2$

307. $(2wx^2y^3)^3$

308. $(^-a^3bc)^6 + 5a^{18}(b^2c^2)^3$

309. $\frac{8xy^2}{2xy}$

310. $(4a^2)^3 + (2a^3)^2 - 11a^6$

311. $\frac{(3x)^3}{x^2 \bullet x^4}$

312. $\frac{(12s^2)(2s^4)}{3s^3}$

313. $\frac{7a^3b^5}{28ab^2}$

314. $(3xy^5)^2 - 11x^2y^2(4y^4)^2$

315. $\frac{2(3x^2y)^2(xy)^3}{3(xy)^2}$

316. $\frac{18x^3y^{-2}z^{-1}}{6x^{-2}y^3z^2}$

317. $4x^{-2}(3ax)^3$

318. $\frac{3x^{-2}}{x^5} - \frac{2x}{x^8}$

319. $\frac{(-2t^{-2}s^4u^{-1})^3(4t^3su^{-2})^2}{8(ts^2u^3)^{-2}}$

320. $\frac{24x^3}{(2x)^2} + \frac{3x^5}{x^4} - \frac{(3ax)^2}{a^2x}$

321. $(4x^2y)^3 + \frac{(2x^2y)^4}{2x^2y}$

322. $(ab^2)^3 + 2b^2 - (4a)^3b^6$

323. $\frac{(4b)^2x^{-2}}{(2ab^2x)^2}$

324. $\frac{ts^2(2t^2s^{-1})^{-2}}{4^{-2}s^{-1}t^{-2}} - \frac{3(t^{-3}s^2)^{-2}}{t(s^{-3}t^2)^3}$

325. $(2xy \cdot \frac{4}{x})^2 + \frac{9y^2}{(3y)^2}$

Answers

Numerical expressions in parentheses like this [ ] are operations performed on only part of the original expression. The operations performed within these symbols are intended to show how to evaluate the various terms that make up the entire expression.

Expressions with parentheses that look like this ( ) contain either numerical substitutions or expressions that are part of a numerical expression. Once a single number appears within these parentheses, the parentheses are no longer needed and need not be used the next time the entire expression is written.

When two pair of parentheses appear side by side like this ( )( ), it means that the expressions within are to be multiplied.

Sometimes parentheses appear within other parentheses in numerical or algebraic expressions. Regardless of what symbol is used, ( ), { }, or [ ], perform operations in the innermost parentheses first and work outward.

Underlined expressions show the simplified result.

301. Add like terms. $\underline{5x^2 + 8x^2 = 13x^2}$

302. Subtract like terms. $\underline{5ab^4 - ab^4 = 4ab^4}$

303.

Use the commutative property of addition.	$9mn^3 + 2mn^3 + 8mn$
Combine like terms.	$\underline{11mn^3 + 8mn}$

304.

Use the commutative property of addition.	$5c^2 - 2c^2 - 7c + 3c + 4$
Combine like terms.	$\underline{3c^2 - 4c + 4}$

305.

Use the commutative property of addition.	$3x^2 + 7x^2 + 4ax - 2ax - 8a^2 + 7a^2$
Combine like terms.	$\underline{10x^2 + 2ax - a^2}$

306.

Use the commutative property of multiplication.	$5 \cdot 6xxyy + 7x^2y^2$
When the same variables are multiplied, add the exponents of the variables.	$(30x^2y^2) + 7x^2y^2$
Combine like terms.	$\underline{37x^2y^2}$

307.

The term within the parentheses is the base of the exponent outside the parentheses.	$(2wx^2y^3)(2wx^2y^3)(2wx^2y^3)$
Use the distributive property of multiplication.	$(2 \cdot 2 \cdot 2)(w \cdot w \cdot w)(x^2 \cdot x^2 \cdot x^2)(y^3 \cdot y^3 \cdot y^3)$
When the same variables are multiplied, add the exponents of the variables.	$(8)(w^{1+1+1})(x^{2+2+2})(y^{3+3+3})$
Simplify.	$\underline{8w^3x^6y^9}$

308. Multiply the exponents of each factor inside the parentheses by the exponent outside the parentheses. $(^-1)^6(a^{18}b^6c^6) + 5a^{18}b^6c^6$

Evaluate the numerical coefficient. $a^{18}b^6c^6 + 5a^{18}b^6c^6$

Combine like terms. $\underline{6a^{18}b^6c^6}$

309. Divide numerical terms. $\frac{8xy^2}{2xy} = \frac{4xy^2}{xy}$

When similar factors, or bases, are being divided, subtract the exponent in the denominator from the exponent in the numerator. $\frac{4xy^2}{xy} = 4x^{1-1}y^{2-1}$

Simplify. $4x^0y^1 = 4(1)y = \underline{4y}$

310. Terms within parentheses are the bases of the exponent outside the parentheses. $(4a^2)(4a^2)(4a^2) + (2a^3)(2a^3) - 11a^6$

Use the distributive property of multiplication. $(4 \cdot 4 \cdot 4)(a^2 \cdot a^2 \cdot a^2) + (2 \cdot 2)(a^3 \cdot a^3) - 11a^6$

When the same variables are multiplied, add the exponents of the variables. $64(a^{2+2+2}) + 4(a^{3+3}) - 11a^6$

Simplify. $64a^6 + 4a^6 - 11a^6$

Combine like terms. $\underline{57a^6}$

Another way of solving this problem is to multiply the exponents of each factor inside the parentheses by the exponent outside of the parentheses. $4^{(1)3}a^{(2)3} + 2^{(1)2}a^{(3)2} - 11a^6$

Simplify the expressions in the exponents. $4^3a^6 + 2^2a^6 - 11a^6$

Simplify terms. $64a^6 + 4a^6 - 11a^6$

Combine like terms. $\underline{57a^6}$

311. In the numerator, multiply the exponents of each factor inside the parentheses by the exponent outside of the parentheses. $\frac{3^3x^3}{x^2 \cdot x^4}$

In the denominator, add the exponents of similar bases. $\frac{3^3x^3}{x^{2+4}} = \frac{27x^3}{x^6}$

When similar factors, or bases, are being divided, you subtract the exponent in the denominator from the exponent in the numerator. $27x^{3-6} = 27x^{-3}$

A base with a negative exponent in the numerator is equivalent to the same variable or base in the denominator with the inverse sign for the exponent. $\underline{27x^{-3} = \frac{27}{x^3}}$

312. Commutative property of multiplication. $\frac{12 \cdot 2s^2 \cdot s^4}{3s^3}$

When similar factors, or bases, are multiplied, add the exponents of the variables. $\frac{12 \cdot 2s^{2+4}}{3s^3}$

Simplify the exponent and coefficients. $\frac{8s^6}{s^3}$

When similar factors, or bases, are being divided, subtract the exponent in the denominator from the exponent in the numerator. $8s^{6-3} = 8s^3$

313. Divide out the common factor of 7 in the numerator and denominator. $\frac{1 \cdot 7a^3b^5}{4 \cdot 7ab^2} = \frac{1a^3b^5}{4ab^2}$

When similar factors, or bases, are being divided, subtract the exponent in the denominator from the exponent in the numerator. $\frac{1a^3b^5}{4ab^2} = \frac{1a^{3-1}b^{5-2}}{4}$

Simplify exponents. $\frac{1a^2b^3}{4}$

Separate the coefficient from the variable. $\frac{1a^2b^3}{4} = \frac{1}{4}a^2b^3$

Both expressions are suitably simplified.

314. Multiply the exponents of each factor inside the parentheses by the exponent outside the parentheses. $3^2x^2y^{10} - 11x^2y^24^2y^8$

Use the commutative property of multiplication. $3^2x^2y^{10} - 11 \cdot 4^2x^2y^2y^8$

When similar factors, or bases, are multiplied, add the exponents of the variables. $3^2x^2y^{10} - 11 \cdot 16 \cdot x^2y^{10}$

Evaluate numerical factors. $9x^2y^{10} - 176x^2y^{10}$

Combine like terms. $^-167x^2y^{10}$

315. Multiply the exponents of each factor inside the parentheses by the exponent outside the parentheses. $\frac{2(3^2x^4y^2)(x^3y^3)}{3(x^2y^2)}$

Use the commutative property of multiplication. $\frac{2 \cdot 9 \cdot x^4x^3y^2y^3}{3x^2y^2}$

When similar factors, or bases, are multiplied, add the exponents of the variables. $\frac{2 \cdot 9 \cdot x^{4+3}y^{2+3}}{3x^2y^2}$

Factor out like numerical terms in the fraction, and simplify exponents with operations. $\frac{6x^7y^5}{x^2y^2}$

When similar factors, or bases, are being divided, subtract the exponent in the denominator from the exponent in the numerator. $6x^{7-2}y^{5-2}$

Simplify the operations in the exponents. $6x^5y^3$

316. When similar factors, or bases, are being divided, subtract the exponent in the denominator from the exponent in the numerator. $\left(\frac{18}{6}\right)x^{3-(-2)}y^{-2-3}z^{-1-2}$

Evaluate the numerical coefficient. $3x^{3-(^-2)}y^{^-2-3}z^{^-1-2}$

Simplify the operations in the exponents. $3x^5y^{^-5}z^{^-3}$

A base with a negative exponent in the numerator is equivalent to the same variable or base in the denominator with the inverse sign for the exponent. $\frac{3x^5}{y^5z^3}$

317. Multiply the exponents of each factor inside the parentheses by the exponent outside the parentheses. $4x^{-2}(3^3a^3x^3)$

Use the commutative property of multiplication. $4 \cdot 3^3 \cdot a^3x^3x^{-2}$

Evaluate numerical terms. $108a^3x^3x^{-2}$

When similar factors, or bases, are being multiplied, add the exponents. $108a^3x^{3-2}$

Simplify the operations in the exponents. $\underline{108a^3x}$

318. When similar factors, or bases, are being divided, subtract the exponent in the denominator from the exponent in the numerator. $3x^{-2-5} - 2x^{1-8}$

Simplify the operations in the exponents. $3x^{-7} - 2x^{-7}$

Subtract like terms. x^{-7}

A base with a negative exponent in the numerator is equivalent to the same variable or base in the denominator with the inverse sign for the exponent. $\underline{x^{-7} = \frac{1}{x^7}}$

319. Multiply the exponents of each factor inside the parentheses by the exponent outside the parentheses.

$$\frac{((-2)^3t^{-6}s^{12}u^{-3})(4^2t^6s^2u^{-4})}{8(t^{-2}s^{-4}u^{-6})}$$

Use the commutative law of multiplication.

$$\frac{(-2)^34^2t^{-6}t^6s^{12}s^2u^{-3}u^{-4}}{8t^{-2}s^{-4}u^{-6}}$$

When the same variables are multiplied, add the exponents of the variables.

$$\frac{-8 \cdot 16t^{-6+6}s^{12+2}u^{-3-4}}{8t^{-2}s^{-4}u^{-6}}$$

Simplify the operations in the exponents.

$$\frac{-8 \cdot 16t^0s^{14}u^{-7}}{8t^{-2}s^{-4}u^{-6}}$$

Factor out like numerical terms in the fraction, and simplify exponents with operations.

$$\frac{-16t^0s^{14}u^{-7}}{t^{-2}s^{-4}u^{-6}}$$

When similar factors, or bases, are being divided, subtract the exponent in the denominator from the exponent in the numerator.

$$-16t^{0-(-2)}s^{14-(-4)}u^{-7-(-6)}$$

Simplify the operations in the exponents.

$$-16t^2s^{18}u^{-1} - 16t^2s^{18}u^{-1}$$

A base with a negative exponent in the numerator is equivalent to the same variable or base in the denominator with the inverse sign for the exponent.

$$\underline{\frac{-16t^2s^{18}}{u}}$$

320. Multiply the exponents of each factor inside the parentheses by the exponent outside the parentheses. $\frac{24x^3}{2^2x^2} + \frac{3x^5}{x^4} - \frac{3^2a^2x^2}{a^2x}$

Evaluate the numerical coefficients and divide out common numerical factors in the terms. $\frac{6x^3}{x^2} + \frac{3x^5}{x^4} - \frac{9a^2x^2}{a^2x}$

When similar factors, or bases, are being divided, subtract the exponent in the denominator from the exponent in the numerator. $6x^{3-2} + 3x^{5-4} - 9a^{2-2}x^{2-1}$

Simplify the operations in the exponents. $6x^1 + 3x^1 - 9a^0x^1$

Any term to the power of zero equals 1. $6x + 3x - 9(1)x$

Combine like terms. $\underline{6x + 3x - 9x = 0x = 0}$

321. Multiply the exponents of each factor inside the parentheses by the exponent outside the parentheses. $4^3x^6y^3 + \frac{2^4x^8y^4}{2x^2y}$

When similar factors, or bases, are being divided, subtract the exponent in the denominator from the exponent in the numerator. $4^3x^6y^3 + 2^{4-1}x^{8-2}y^{4-1}$

Simplify the operations in the exponents. $4^3x^6y^3 + 2^3x^6y^3$

Evaluate the numerical coefficients. $64x^6y^3 + 8x^6y^3$

Add like terms. $\underline{72x^6y^3}$

322. Multiply the exponents of each factor inside the parentheses by the exponent outside the parentheses. $a^3b^6 + 2b^2 - 4^3a^3b^6$

Evaluate the numerical coefficients. $a^3b^6 + 2b^2 - 64a^3b^6$

Use the commutative property of addition. $2b^2 - 64a^3b^6 + a^3b^6$

Combine like terms in the expression. $\underline{2b^2 - 63a^3b^6}$

323. Multiply the exponents of each factor inside the parentheses by the exponent outside the parentheses. $\frac{4^2b^2x^{-2}}{2^2a^2b^4x^2}$

Evaluate the numerical coefficients. $\frac{16b^2x^{-2}}{4a^2b^4x^2}$

Simplify the numerical factors in the numerator and the denominator. $\frac{4b^2x^{-2}}{a^2b^4x^2}$

When similar factors, or bases, are being divided, subtract the exponent in the denominator from the exponent in the numerator. $\frac{4b^{2-4}x^{-2-2}}{a^2}$

Simplify the operations in the exponents. $\frac{4b^{-2}x^{-4}}{a^2}$

A base with a negative exponent in the numerator is equivalent to the same variable or base in the denominator with the inverse sign for the exponent. $\underline{\frac{4}{a^2b^2x^4}}$

324. Multiply the exponents of each factor inside the parentheses by the exponent outside the parentheses. $\frac{ts^2(2^{-2}t^{-4}s^2)}{4^{-2}s^{-1}t^{-2}} - \frac{3(t^6s^{-4})}{t(s^{-9}t^6)}$

Use the commutative law of multiplication. $\frac{2^{-2}t^{-4}ts^2s^2}{4^{-2}s^{-1}t^{-2}} - \frac{3t^6s^{-4}}{s^{-9}t^6t}$

When the same variables are multiplied, add the exponents of the variables. $\frac{2^{-2}t^{-4+1}s^{2+2}}{4^{-2}s^{-1}t^{-2}} - \frac{3t^6s^{-4}}{s^{-9}t^{6+1}}$

Simplify the operations in the exponents. $\frac{2^{-2}t^{-3}s^4}{4^{-2}s^{-1}t^{-2}} - \frac{3t^6s^{-4}}{s^{-9}t^7}$

When similar factors, or bases, are being divided, subtract the exponent in the denominator from the exponent in the numerator. $\frac{2^{-2}t^{-3-(-2)}s^{4-(-1)}}{4^{-2}} - \frac{3t^{6-7}s^{-4-(-9)}}{1}$

Simplify the operations in the exponents. $\frac{2^{-2}t^{-1}s^5}{4^{-2}} - \frac{3t^{-1}s^5}{1}$

A base with a negative exponent in the numerator (or denominator) is equivalent to the same variable or base in the denominator (or numerator) with the inverse sign for the exponent. $\frac{4^2s^5}{2^2t} - \frac{3s^5}{t}$

Simplify the numerical coefficients. $4\left(\frac{s^5}{t}\right) - 3\left(\frac{s^5}{t}\right)$

Add like terms. $\underline{\frac{s^5}{t}}$

325. Multiply the exponents of each factor inside the parentheses by the exponent outside the parentheses. $(2xy)^2 \cdot (\frac{4}{x})^2 + \frac{9y^2}{3^2y^2}$

Repeat the previous step. $(2^2x^2y^2)(\frac{4^2}{x^2}) + \frac{9y^2}{3^2y^2}$

Evaluate numerical factors. $(4x^2y^2)(\frac{16}{x^2}) + \frac{9y^2}{9y^2}$

The last term is equivalent to 1. $(4x^2y^2)(\frac{16}{x^2}) + 1$

Multiply the fraction in the first term by the factor in the first term. $(\frac{(4x^2y^2)16}{x^2}) + 1$

Use the commutative property of multiplication. $(\frac{4 \cdot 16x^2y^2}{x^2}) + 1$

Evaluate numerical factors. $\frac{64x^2y^2}{x^2} + 1$

Divide out the common factor of x^2 in the numerator and denominator. $\underline{64y^2 + 1}$

14

Multiplying Polynomials

This chapter will present problems for you to solve in the multiplication of polynomials. Specifically, you will practice solving problems multiplying a monomial (one term) and a polynomial, multiplying binomials (expressions with two terms), and multiplying a trinomial and a binomial.

Tips for Multiplying Polynomials

When multiplying a polynomial by a monomial, use the distributive property of multiplication to multiply each term in the polynomial by the monomial.

$$a(b + c + d + e) = ab + ac + ad + ae$$

When multiplying a binomial by a binomial, use the mnemonic FOIL to remind you of the different pairs of terms you must multiply together. $(a + b)(c + d)$

F is for **first**. Multiply the first terms of each binomial. $([a] + b)([c] + d)$ gives the term ac.

O is for **outer**. Multiply the outer terms of each binomial. $([a] + b)(c + [d])$ gives the term ad.

I is for **inner**. Multiply the inner terms of each binomial. $(a + [b])([c] + d)$ gives the term bc.

L is for **last**. Multiply the last terms of each binomial. $(a + [b])(c + [d])$ gives the term bd.

Then you add the four terms. $ac + ad + bc + bd$

Multiplying a trinomial by a binomial is done similarly. You proceed as you would when using the distributive property of multiplication. Multiply each term in the trinomial by the first and then the second term in the binomial. Then add the various terms.

$$(a + b)(c + d + e) = (ac + ad + ae) + (bc + bd + be)$$

Multiply the following polynomials.

326. $x(5x + 3y - 7)$

327. $2a(5a^2 - 7a + 9)$

328. $4bc(3b^2c + 7b - 9c + 2bc^2 - 8)$

329. $3mn(^-4m + 6n + 7mn^2 - 3m^2n)$

330. $^-5x^2(6x^5y - 2x^3y^2 + 4xy^3 - 5)$

331. $(x + 3)(x + 6)$

332. $(x - 4)(x - 9)$

333. $(2x + 1)(3x - 7)$

334. $(3 - 2x)(1 - 5x)$

335. $(7x + 2y)(2x - 4y)$

336. $(5x + 7)(5x - 7)$

337. $^-3\left(\frac{2}{3} - \frac{1}{2}x\right)(18 - 6x)$

338. $(3x^2 + y^2)(x^2 - 2y^2)$

339. $(4 + 2x^2)(9 - 3x)$

340. $(5x^3 - 2x)(3 - 4x^2)$

341. $(x + 2)(3x^2 - 5x + 2)$

342. $(2x - 3)(x^3 + 3x^2 - 4x)$

343. $(4a + b)(5a^2 + 2ab - b^2)$

344. $(3y - 7)(6y^2 - 3y + 7)$

345. $(3 - 2x)(1 - x - 3x^2)$

346. $(x + 2)(2x + 1)(x - 1)$

347. $(3a - 4)(5a + 2)(a + 3)$

348. $(2n - 3)(2n + 3)(n + 4)$

349. $(5r - 7)(3r^4 + 2r^2 + 6)$

350. $(3x^2 + 4)(x - 3)(3x^2 - 4)$

Answers

Numerical expressions in parentheses like this [ ] are operations performed on only part of the original expression. The operations performed within these symbols are intended to show how to evaluate the various terms that make up the entire expression.

Expressions with parentheses that look like this ( ) contain either numerical substitutions or expressions that are part of a numerical expression. Once a single number appears within these parentheses, the parentheses are no longer needed and need not be used the next time the entire expression is written.

When two pair of parentheses appear side by side like this ( )( ), it means that the expressions within are to be multiplied.

Sometimes parentheses appear within other parentheses in numerical or algebraic expressions. Regardless of what symbol is used, ( ), { }, or [ ], perform operations in the innermost parentheses first and work outward.

Underlined expressions show the simplified result.

326. Multiply each term in the trinomial by x. $x(5x) + x(3y) - x(7)$
Simplify terms. $\underline{5x^2 + 3xy - 7x}$

327. Multiply each term in the trinomial by $2a$. $2a(5a^2) - 2a(7a) + 2a(9)$
Simplify terms. $\underline{10a^3 - 14a^2 + 18a}$

328. Multiply each term in the polynomial by $4bc$. $4bc(3b^2c) + 4bc(7b) - 4bc(9c) + 4bc(2bc^2) - 4bc(8)$
Simplify terms. $\underline{12b^3c^2 + 28b^2c - 36bc^2 + 8b^2c^3 - 32bc}$

329. Multiply each term in the polynomial by $3mn$. $3mn(^-4m) + 3mn(6n) + 3mn(7mn^2) - 3mn(3m^2n)$
Simplify terms. $\underline{^-12m^2n + 18mn^2 + 21m^2n^3 - 9m^3n^2}$

330. Multiply each term in the polynomial by $^-5x^2$. $^-5x^2(6x^5y) - 5x^2(^-2x^3y^2) - 5x^2(4xy^3) - 5x^2(^-5)$
Simplify terms. $\underline{^-30x^7y + 10x^5y^2 - 20x^3y^3 + 25x^2}$

331. Use **FOIL** to multiply binomials.
Multiply the **first** terms in each binomial. $([x] + 3)([x] + 6)$
x^2
Multiply the **outer** terms in each binomial. $([x] + 3)(x + [6])$
^+6x
Multiply the **inner** terms in each binomial. $(x + [3])([x] + 6)$
^+3x
Multiply the **last** terms in each binomial. $(x + [3])(x + [6])$
$^+18$
Add the products of FOIL together. $x^2 + 6x + 3x + 18$
Combine like terms. $\underline{x^2 + 9x + 18}$

332. Use **FOIL** to multiply binomials.

Multiply the **first** terms in each binomial.	$([x] - 4)([x] - 9)$ x^2
Multiply the **outer** terms in each binomial.	$([x] - 4)(x - [9])$ ^-9x
Multiply the **inner** terms in each binomial.	$(x - [4])([x] - 9)$ ^-4x
Multiply the **last** terms in each binomial.	$(x - [4])(x - [9])$ $^+36$
Add the products of FOIL together.	$x^2 - 9x - 4x + 36$
Combine like terms.	$\underline{x^2 - 13x + 36}$

333. Use **FOIL** to multiply binomials.

Multiply the **first** terms in each binomial.	$([2x] + 1)([3x] - 7)$ $6x^2$
Multiply the **outer** terms in each binomial.	$([2x] + 1)(3x - [7])$ ^-14x
Multiply the **inner** terms in each binomial.	$(2x + [1])([3x] - 7)$ ^+3x
Multiply the **last** terms in each binomial.	$(2x + [1])(3x - [7])$ $^-7$
Add the products of FOIL together.	$6x^2 - 14x + 3x - 7$
Combine like terms.	$\underline{6x^2 - 11x - 7}$

334. Use **FOIL** to multiply binomials.

Multiply the **first** terms in each binomial.	$([3] - 2x)([1] - 5x)$ 3
Multiply the **outer** terms in each binomial.	$([3] - 2x)(1 - [5x])$ ^-15x
Multiply the **inner** terms in each binomial.	$(3 - [2x])([1] - 5x)$ ^-2x
Multiply the **last** terms in each binomial.	$(3 - [2x])(1 - [5x])$ $10x^2$
Add the products of FOIL together.	$10x^2 - 2x - 15x + 3$
Combine like terms.	$\underline{10x^2 - 17x + 3}$

335. Use **FOIL** to multiply binomials.

Multiply the **first** terms in each binomial.	$([7x] + 2y)([2x] - 4y)$ $14x^2$
Multiply the **outer** terms in each binomial.	$([7x] + 2y)(2x - [4y])$ ^-28xy
Multiply the **inner** terms in each binomial.	$(7x + [2y])([2x] - 4y)$ ^+4xy

Multiply the **last** terms in each binomial. $(7x + [2y])(2x - [4y])$
$^{-}8y^2$

Add the products of FOIL together. $14x^2 - 28xy + 4xy - 8y^2$
Combine like terms. $\underline{14x^2 - 24xy - 8y^2}$

336. Use **FOIL** to multiply binomials.
Multiply the **first** terms in each binomial. $([5x] + 7)([5x] - 7)$
$25x^2$

Multiply the **outer** terms in each binomial. $([5x] + 7)(5x - [7])$
$^{-}35x$

Multiply the **inner** terms in each binomial. $(5x + [7])([5x] - 7)$
$^{+}35x$

Multiply the **last** terms in each binomial. $(5x + [7] (5x - [7])$
$^{-}49$

Add the products of FOIL together. $25x^2 - 35x + 35x - 49$
Combine like terms. $\underline{25x^2 - 49}$

337. First, use the distributive law of multiplication to combine the first two terms.
$\left[^{-}3\left(\frac{2}{3}\right) - 3\left(^{-}\frac{1}{2}x\right)\right](18 - 6x)$
$\left(^{-}2 + \frac{3}{2}x\right)(18 - 6x)$

Use **FOIL** to multiply binomials.
Multiply the **first** terms in each binomial. $\left([^{-}2] + \frac{3}{2}x\right)(18 - [6x])$
$12x$

Multiply the **inner** terms in each binomial. $\left(^{-}2 + \left[\frac{3}{2}x\right]\right)([18] - 6x)$
$27x$

Multiply the **last** terms in each binomial $\left(^{-}2 + \left[\frac{3}{2}x\right]\right)(18 - [6x])$
$^{-}9x^2$

Add the products of FOIL together. $^{-}9x^2 + 27x + 12x - 36$

Combine like terms. $\underline{^{-}9x^2 + 39x - 36}$

338. Use **FOIL** to multiply binomials.
Multiply the **first** terms in each binomial. $([3x^2] + y^2)([x^2] - 2y^2)$
$3x^4$

Multiply the **outer** terms in each binomial. $([3x^2] + y^2)(x^2 - [2y^2])$
$^{-}6x^2y^2$

Multiply the **inner** terms in each binomial. $(3x^2 + [y^2])([x^2] - 2y^2$
$^{+}x^2y^2$

Multiply the **last** terms in each binomial. $(3x^2 + [y^2])(x^2 - [2y^2])$
$^{-}2y^4$

Add the products of FOIL together. $3x^4 - 6x^2y^2 + x^2y^2 - 2y^4$

Combine like terms. $\underline{3x^4 - 5x^2y^2 - 2y^4}$

339. Use **FOIL** to multiply binomials.
Multiply the **first** terms in each binomial. $([4] + 2x^2)([9] - 3x)$
$^{+}36$

Multiply the **outer** terms in each binomial. $([4] + 2x^2)(9 - [3x])$
$^{-}12x$

Multiply the **inner** terms in each binomial. $(4 + [2x^2])([9] - 3x)$
$^{+}18x^2$

Multiply the **last** terms in each binomial. $(4 + [2x^2])(9 - [3x])$
$^{-}6x^3$

Add the products of FOIL together. $36 - 12x + 18x^2 - {}^{-}6x^3$
Simplify and put them in order from the highest power. $\underline{{}^{-}6x^3 + 18x^2 - 12x + 36}$

340. Use **FOIL** to multiply binomials.
Multiply the **first** terms in each binomial. $([5x^3] - 2x)([3] - 4x^2)$
$15x^3$

Multiply the **outer** terms in each binomial. $([5x^3] - 2x)(3 - [4x^2])$
$^{-}20x^5$

Multiply the **inner** terms in each binomial. $(5x^3 - [2x])([3] - 4x^2)$
$^{-}6x$

Multiply the **last** terms in each binomial. $(5x^3 - [2x])(3 - [4x^2])$
$8x^3$

Add the products of FOIL together. $8x^3 - 6x - 20x^5 + 15x^3$

Combine like terms. $\underline{{}^{-}20x^5 + 23x^3 - 6x}$

341. Multiply the trinomial by the first term in the binomial, x. $[x(3x^2 - 5x + 2)]$
$[x(3x^2) - x(5x) + x(2)]$
Simplify terms. $[3x^3 - 5x^2 + 2x]$
Multiply the trinomial by the second term in the binomial, 2. $[2(3x^2 - 5x + 2)]$
$[2(3x^2) - 2(5x) + 2(2)]$
Simplify terms. $[6x^2 - 10x + 4]$
Add the results of multiplying by the terms in the binomial together. $[3x^3 - 5x^2 + 2x] + [6x^2 - 10x + 4]$
Use the commutative property of addition. $3x^3 - 5x^2 + 6x^2 + 2x - 10x + 4$
Combine like terms. $\underline{3x^3 + x^2 - 8x + 4}$

342. Multiply the trinomial by the first term in the binomial, $2x$. $[2x(x^3 + 3x^2 - 4x)]$
Use the distributive property of multiplication. $[2x(x^3) + 2x(3x^2) - 2x(4x)]$
Simplify terms. $[2x^4 + 6x^3 - 8x^2]$

Multiply the trinomial by the second term in the binomial, $^-3$. $[^-3(x^3 + 3x^2 - 4x)]$
Use the distributive property of multiplication. $[^-3(x^3) - 3(3x^2) - 3(^-4x)]$
Simplify terms. $[^-3x^3 - 9x^2 + 12x]$
Add the results of multiplying by the terms in the binomial together. $[2x^4 + 6x^3 - 8x^2] + [^-3x^3 - 9x^2 + 12x]$
Use the commutative property of addition. $2x^4 + 6x^3 - 3x^3 - 8x^2 - 9x^2 + 12x$
Combine like terms. $\underline{2x^4 + 3x^3 - 17x^2 + 12x}$

343. Multiply the trinomial by the first term in the binomial, $4a$. $[4a(5a^2 + 2ab - b^2)]$
Use the distributive property of multiplication. $[4a(5a^2) + 4a(2ab) - 4a(b^2)]$
Simplify terms. $[20a^3 + 8a^2b - 4ab^2]$
Multiply the trinomial by the second term in the binomial, b. $[b(5a^2 + 2ab - b^2)]$
Use the distributive property of multiplication. $[b(5a^2) + b(2ab) - b(b^2)]$
Simplify terms. $[5a^2b + 2ab^2 - b^3]$
Add the results of multiplying by the terms in the binomial together. $[20a^3 + 8a^2b - 4ab^2] + [5a^2b + 2ab^2 - b^3]$
Use the commutative property of addition. $20a^3 + 8a^2b + 5a^2b - 4ab^2 + 2ab^2 - b^3$
Combine like terms. $\underline{20a^3 + 13a^2b - 2ab^2 - b^3}$

344. Multiply the trinomial by the first term in the binomial, $3y$. $[3y(6y^2 - 3y + 7)]$
Use the distributive property of multiplication. $[3y(6y^2) - 3y(3y) + 3y(7)]$
Simplify terms. $[18y^3 - 9y^2 + 21y]$
Multiply the trinomial by the second term in the binomial, $^-7$. $[^-7(6y^2 - 3y + 7)]$
Use the distributive property of multiplication. $[^-7(6y^2) - 7(^-3y) - 7(7)]$
Simplify terms. $[^-42y^2 + 21y - 49]$
Add the results of multiplying by the terms in the binomial together. $[18y^3 - 9y^2 + 21y] + [^-42y^2 + 21y - 49]$
Use the commutative property of addition. $18y^3 - 9y^2 - 42y^2 + 21y + 21y - 49$
Combine like terms. $\underline{18y^3 - 51y^2 + 42y - 49}$

345. Multiply the trinomial by the first term in the binomial, 3. $3(1 - x - 3x^2)$

Use the distributive property of multiplication. $3(1) + 3(^-x) + 3(^-3x^2)$

Simplify terms. $3 - 3x - 9x^2$

Multiply the trinomial by the second term in the binomial, ^-2x. $(^-2x)(1 - x - 3x^2)$

Use the distributive property of multiplication. $(^-2x)(1) + (^-2x)(^-x) + (^-2x)(^-3x^2)$

Simplify terms. $^-2x + 2x^2 + 6x^3$

Add the results of multiplying by the terms in the binomial together. $3 - 3x - 9x^2 - 2x + 2x^2 + 6x^3$

Use the commutative property of addition. $3 - 3x - 2x - 9x^2 + 2x^2 + 6x^3$

Combine like terms. $3 - 5x - 7x^2 + 6x^3$

Use the commutative property of addition. $\underline{6x^3 - 7x^2 - 5x + 3}$

346. Multiply the first two binomials in the expression using **FOIL**.

Multiply the **first** terms in each binomial. $([x] + 2)([2x] + 1)$
$2x^2$

Multiply the **outer** terms in each binomial. $([x] + 2)(2x + [1])$
^+x

Multiply the **inner** terms in each binomial. $(x + [2])([2x] + 1)$
^+4x

Multiply the **last** terms in each binomial. $(x + [2])(2x + [1])$
$^+2$

Add the products of FOIL together. $2x^2 + x + 4x + 2$

Combine like terms. $2x^2 + 5x + 2$

Multiply the resulting trinomial by the last binomial in the original expression. $(x - 1)(2x^2 + 5x + 2)$

Multiply the trinomial by the first term in the binomial, x. $[x(2x^2 + 5x + 2)]$

Use the distributive property of multiplication. $[x(2x^2) + x(5x) + x(2)]$

Simplify terms. $[2x^3 + 5x^2 + 2x]$

Multiply the trinomial by the second term in the binomial, $^-1$. $[^-1(2x^2 + 5x + 2)]$

Use the distributive property of multiplication. $[^-2x^2 - 5x - 2]$

Add the results of multiplying by the terms in the binomial together.	$[2x^3 + 5x^2 + 2x] + [^-2x^2 - 5x - 2]$
Use the commutative property of addition.	$2x^3 + 5x^2 - 2x^2 + 2x - 5x - 2$
Combine like terms.	$\underline{2x^3 + 3x^2 - 3x - 2}$

347.

Multiply the first two binomials in the expression using **FOIL**.	
Multiply the **first** terms in each binomial.	$([3a] - 4)([5a] + 2)$ $15a^2$
Multiply the **outer** terms in each binomial.	$([3a] - 4)(5a + [2])$ ^+6a
Multiply the **inner** terms in each binomial.	$(3a - [4])([5a] + 2)$ ^-20a
Multiply the **last** terms in each binomial.	$(3a - [4])(5a + [2])$ $^-8$
Add the products of FOIL together.	$15a^2 + 6a - 20a - 8$
Combine like terms.	$15a^2 - 14a - 8$
Multiply the resulting trinomial by the last binomial in the original expression.	$(a + 3)(15a^2 - 14a - 8)$
Multiply the trinomial by the first term in the binomial, *a*.	$[a(15a^2 - 14a - 8)]$
Use the distributive property of multiplication.	$[a(15a^2) - a(14a) - a(8)]$
Simplify terms.	$[15a^3 - 14a^2 - 8a]$
Multiply the trinomial by the second term in the binomial, 3.	$[3(15a^2) - 3(14a) - 3(8)]$
Use the distributive property of multiplication.	$[45a^2 - 42a - 24]$
Add the results of multiplying by the terms in the binomial together.	$[15a^3 - 14a^2 - 8a] + [45a^2 - 42a - 24]$
Use the commutative property of addition.	$15a^3 - 14a^2 + 45a^2 - 8a - 42a - 24$
Combine like terms.	$\underline{15a^3 + 31a^2 - 50a - 24}$

348. Multiply the first two binomials in the expression using **FOIL**.	
Multiply the **first** terms in each binomial.	$([2n] - 3)([2n] + 3)$ $4n^2$
Multiply the **outer** terms in each binomial.	$([2n] - 3)(2n + [3])$ $^{+}6n$
Multiply the **inner** terms in each binomial.	$(2n - [3])([2n] + 3)$ $^{-}6n$
Multiply the **last** terms in each binomial.	$(2n - [3])(2n + [3])$ $^{-}9$
Add the products of FOIL together.	$4n^2 + 6n - 6n - 9$
Combine like terms.	$4n^2 - 9$
Now we again have two binomials. Use FOIL to find the solution.	$(n + 4)(4n^2 - 9)$
Multiply the **first** terms in each binomial.	$([n] + 4)([4n^2] - 9)$ $4n^3$
Multiply the **outer** terms in each binomial.	$([n] + 4)(4n^2 - [9])$ $^{-}9n$
Multiply the **inner** terms in each binomial.	$(n + [4])([4n^2] - 9)$ $^{+}16n^2$
Multiply the **last** terms in each binomial.	$(n + [4])(4n^2 - [9])$ $^{-}36$
Add the products of FOIL together.	$4n^3 - 9n + 16n^2 - 36$
Order terms from the highest to lowest power.	$\underline{4n^3 + 16n^2 - 9n - 36}$

349. Multiply the trinomial by the first term in the binomial, $5r$.	$[5r(3r^4 + 2r^2 + 6)]$
Use the distributive property of multiplication.	$[5r(3r^4) + 5r(2r^2) + 5r(6)]$
Simplify terms.	$[15r^5 + 10r^3 + 30r]$
Multiply the trinomial by the second term in the binomial, $^{-}7$.	$[^{-}7(3r^4 + 2r^2 + 6)]$
Use the distributive property of multiplication.	$[^{-}7(3r^4) - 7(2r^2) - 7(6)]$
Simplify terms.	$[^{-}21r^4 - 14r^2 - 42]$
Add the results of multiplying by the terms in the binomial together.	$[15r^5 + 10r^3 + 30r] + [^{-}21r^4 - 14r^2 - 42]$
Use the commutative property of addition.	$\underline{15r^5 - 21r^4 + 10r^3 - 14r^2 + 30r - 42}$

350. Multiply the first two binomials in the expression using **FOIL**.

Multiply the **first** terms in each binomial.	$([3x^2] + 4)([x] - 3)$ $3x^3$
Multiply the **outer** terms in each binomial.	$([3x^2] + 4)(x - [3])$ $^{-}9x^2$
Multiply the **inner** terms in each binomial.	$(3x^2 + [4])([x] - 3)$ $^{+}4x$
Multiply the **last** terms in each binomial.	$(3x^2 + [4])(x - [3])$ $^{-}12$
Add the products of FOIL together.	$(3x^3 - 9x^2 + 4x - 12)$
Multiply the resulting polynomial by the last binomial in the original expression.	$(3x^2 - 4)(3x^3 - 9x^2 + 4x - 12)$
Multiply the trinomial by the first term in the binomial, $3x^2$.	$[3x^2(3x^3 - 9x^2 + 4x - 12)]$
Use the distributive property of multiplication.	$[3x^2\ (3x^3) - 3x^2(9x^2)$ $+ 3x^2(4x) - 3x^2(12)]$
Simplify terms.	$[9x^5 - 27x^4 + 12x^3 - 36x^2]$
Multiply the polynomial by the second term in the binomial, $^{-}4$.	$[^{-}4(3x^3 - 9x^2 + 4x - 12)]$
Use the distributive property of multiplication.	$[^{-}12x^3 + 36x^2 - 16x + 48]$
Add the results of multiplying by the terms in the binomial together.	$[9x^5 - 27x^4 + 12x^3 - 36x^2]$ $+ [^{-}12x^3 + 36x^2 - 16x + 48]$
Use the commutative property of addition.	$9x^5 - 27x^4 + 12x^3 - 12x^3 -$ $36x^2 + 36x^2 - 16x + 48$
Combine like terms.	$\underline{9x^5 - 27x^4 - 16x + 48}$

15

Factoring Polynomials

This chapter will present algebraic expressions for you to factor. You can use three different techniques to factor polynomials. In the first technique, you look for common factors in the terms of the polynomial. In the second, you will factor polynomials that are the difference of two perfect squares. The third technique, called the trinomial factor method, will allow you to factor algebraic expressions comprised of three terms. The trinomial expressions in this chapter will be in the form of $x^2 \pm ax \pm b$, where a and b are whole numbers. The problems will be presented in random order to give you practice at recognizing which method or combination of methods will be required to factor the polynomial. Complete explanations of the solutions will follow.

Tips for Factoring Polynomials

Factoring using the greatest common factor: Look for a factor common to every term in the polynomial. Put that factor outside a set of parentheses and the polynomial inside with the factor removed from each term, e.g. $2x^2 + 8 = 2(x^2 + 4)$

Factoring using the difference of two perfect squares: Polynomials in the form $x^2 - y^2$ can be factored into a product of two binomials: $(x + y)(x - y)$.

Factoring using the trinomial method: This method requires you to factor the first and third terms and put the factors into the following factored form: ([ ] ± [ ])([ ] ± [ ]). The factors of the first term go in the first position in the parentheses and the factors of the third term go in the second position in each factor, e.g. $x^2 - 7x - 8 = (x - 8)(x + 1)$.

Factor the following polynomials.

351. $9a + 15$

352. $3a^2x + 9ax$

353. $x^2 - 16$

354. $4a^2 - 25$

355. $7n^2 - 28n$

356. $50n^4 - 25n^3$

357. $x^2 + 3x + 2$

358. $9r^2 - 49$

359. $x^2 - 2x - 8$

360. $x^2 + 5x + 6$

361. $x^2 + x - 6$

362. $b^2 - 100$

363. $x^2 + 7x + 12$

364. $x^2 - 20x + 99$

365. $b^2 - 6b + 8$

366. $b^2 + 20b + 84$

367. $a^2 + 11a - 12$

368. $x^2 + 10x + 25$

369. $36y^4 - 4z^2$

370. $c^2 - 12c + 32$

371. $d^3 - 16d^2 + 55d$

372. $w^4 + 17w^3 + 60w^2$

373. $y^5 + 16y^4 + 28y^3$

374. $v^4 - 13v^2 - 48$

375. $x^2 - 20x + 36$

Answers

Numerical expressions in parentheses like this [ ] are operations performed on only part of the original expression. The operations performed within these symbols are intended to show how to evaluate the various terms that make up the entire expression.

Expressions with parentheses that look like this ( ) contain either numerical substitutions or expressions that are part of a numerical expression. Once a single number appears within these parentheses, the parentheses are no longer needed and need not be used the next time the entire expression is written.

When two pair of parentheses appear side by side like this ( )( ), it means that the expressions within are to be multiplied.

Sometimes parentheses appear within other parentheses in numerical or algebraic expressions. Regardless of what symbol is used, ( ), { }, or [ ], perform operations in the innermost parentheses first and work outward.

Underlined expressions show the original algebraic expression as an equation with the expression equal to its factored result.

351. The terms have a common factor of 3. Factor 3 out of each term and write the expression in factored form. $\underline{9a + 15 = 3(3a + 5)}$

352. The terms have a common factor of $3ax$. Factor $3ax$ out of each term and write the expression in factored form. $\underline{3a^2x + 9ax = 3ax(a + 3)}$

353. Both terms in the polynomial are perfect squares. Use the form for factoring the difference of two perfect squares and put the roots of each factor in the proper place. $x^2 - 16 = (x + 4)(x - 4)$

Check using FOIL. $\underline{(x + 4)(x - 4) = x^2 - 4x + 4x - 16 = x^2 - 16}$

354. Both terms in the polynomial are perfect squares. $4a^2 = (2a)^2$, and $25 = 5^2$. Use the form for factoring the difference of two perfect squares and put the roots of each factor in the proper place. $4a^2 - 25 = (2a + 5)(2a - 5)$

Check using FOIL. $\underline{(2a + 5)(2a - 5) = 4a^2 - 10a + 10a - 25 = 4a^2 - 25}$

355. The terms have a common factor of $7n$. Factor $7n$ out of each term and write the expression in factored form. $\underline{7n^2 - 28n = 7n(n - 4)}$

356. The terms have a common factor of $25n^3$. Factor $25n^3$ out of each term in the expression and write the expression in factored form. $\underline{50n^4 - 25n^3 = 25n^3(2n - 1)}$

357. This expression can be factored using the trinomial method. The factors of x^2 are x and x, and the factors of 2 are 1 and 2. Place the factors into the trinomial factor form and check using FOIL. $(x + 2)(x + 1) = x^2 + x + 2x + 2 = x^2 + 3x + 2$

The factors are correct.

358. Both terms in the polynomial are perfect squares. $9r^2 = (3r)^2$ and $49 = 7^2$. Use the form for factoring the difference of two perfect squares and put the roots of each factor in the proper place. $9r^2 - 49 = (3r + 7)(3r - 7)$

359. This expression can be factored using the trinomial method. The factors of x^2 are x and x, and the factors of 8 are (1)(8) and (2)(4). You want the result of the O and I of the FOIL method for multiplying factors to add up to ^-2x. Only terms with opposite signs will result in a negative numerical term, which is what you need, since the third term is $^-8$. Place the factors (2)(4) into the trinomial factor form and check using FOIL. $(x + 4)(x - 2) = x^2 - 2x + 4x - 8 = x^2 + 2x - 8$

Almost correct! Change the position of the factors of the numerical term and check using FOIL. $(x + 2)(x - 4) = x^2 - 4x + 2x - 8 = x^2 - 2x - 8$

The factors of the trinomial are now correct.

360. This expression can be factored using the trinomial method. The factors of x^2 are x and x, and the factors of 6 are (1)(6) and (2)(3). Since the numerical term of the polynomial is positive, the signs in the factor form for trinomials will be the same because only two like signs multiplied together will result in a positive. Now consider the second term in the trinomial. In order to add up to $5x$, the result of multiplying the Inner and Outer terms of the trinomial factors will have to be positive. Try using two positive signs and the factors 2 and 3, which add up to 5. Check using FOIL. $(x + 2)(x + 3) = x^2 + 2x + 3x + 6 = x^2 + 5x + 6$

361. This expression can be factored using the trinomial method. The factors of x^2 are x and x, and the factors of 6 are (1)(6) and (2)(3). You want the result of the O and I of the FOIL method for multiplying factors to add up to $^{+}1x$. Only terms with opposite signs will result in a negative numerical term that you need with the third term being a $^{-}6$. Place the factors (2)(3) into the trinomial factor form and check using FOIL. $\underline{(x + 3)(x - 2) = x^2 - 2x + 3x - 6 = x^2 + x - 6}$
The factors of the trinomial are correct.

362. Both terms in the polynomial are perfect squares. $b^2 = (b)^2$ and $100 = 10^2$. Use the form for factoring the difference of two perfect squares and put the roots of each factor in the proper place. $b^2 - 100 = (b + 10)(b - 10)$
Check using FOIL. $\underline{(b + 10)(b - 10) = b^2 - 10b + 10b - 100 = b^2 - 100}$

363. This expression can be factored using the trinomial method. The factors of x^2 are x and x, and the factors of 12 are (1)(12) or (2)(6) or (3)(4). You want the result of the O and I of the FOIL method for multiplying factors to add up to $^{+}7x$. The factors (3)(4) would give terms that add up to 7. Since all signs are positive, use positive signs in the factored form for the trinomial. $\underline{(x + 3)(x + 4) = x^2 + 3x + 4x + 12 = x^2 + 7x + 12}$
The result is correct. This is not just luck. You can use logical guesses to find the correct combination of factors and signs.

364. This expression can be factored using the trinomial method. The factors of x^2 are x and x, and the factors of 99 are (1)(99), (3)(33), and (9)(11). You want the result of the O and I of the FOIL method for multiplying factors to add up to $-20x$. Only terms with the same sign will result in a positive numerical term, which is what you need, since the third term is 99. Place the factors (9)(11) into the trinomial factor form and check using FOIL. $\underline{(x - 11)(x - 9) = x^2 - 9x - 11x + 99 = x^2 - 20x + 99}$

365. This expression can be factored using the trinomial method. The factors of b^2 are b and b, and the factors of 8 are (1)(8) or (2)(4). You want the result of the O and I of the FOIL method for multiplying factors to add up to ^-6b. The signs within the parentheses of the factorization of the trinomial must be the same to result in a positive numerical term in the trinomial. The middle term has a negative sign, so let's try two negative signs. Use the (2)(4).

Check the answer using FOIL. $\underline{(b-2)(b-4) = b^2 - 4b - 2b + 8 = b^2 - 6b + 8}$

366. This expression can be factored using the trinomial method. The factors of b^2 are b and b, and the factors of 84 are (1)(84), (2)(42), (3)(28), (4)(21), (6)(14), and (7)(12). You want the result of the O and I of the FOIL method for multiplying factors to add up to $20b$. Only terms with the same sign will result in a positive numerical term, which is what you need, since the third term is 84. Place the factors (6)(14) into the trinomial factor form and check using FOIL $\underline{b^2 + 20b + 84 = (b+6)(b+14) = b^2 + 14b + 6b + 84}$

367. This expression can be factored using the trinomial method. The factors of a^2 are a and a, and the factors of 12 are (1)(12) or (2)(6) or (3)(4). You want the result of the O and I of the FOIL method for multiplying factors to add up to ^+11a. Only the product of a positive and a negative numerical term will result in $^-12$. Since the signs in the factors must be one positive and one negative, use the factors 12 and 1 in the trinomial factors form.

Use FOIL to check the answer. $\underline{(a+12)(a-1) = a^2 - 1a + 12a - 12 = a^2 + 11a - 12}$

368. This expression can be factored using the trinomial method. The factors of x^2 are x and x, and the factors of 25 are (1)(25) or (5)(5). To get a positive 25 after multiplying the factors of the trinomial expression, the signs in the two factors must both be positive or both be negative. The sum of the results of multiplying the Outer and Inner terms of the trinomial factors needs to add up to a $^{+}10x$. So let's use (5)(5) in the trinomial factors form and check using FOIL. $\underline{(x + 5)(x + 5) = x^2 + 5x + 5x + 25 = x^2 + 10x + 25}$

369. Both terms in the polynomial are perfect squares. $36y^2 = (6y^2)^2$ and $4z^2 = (2z)^2$
Use the form for factoring the difference of two perfect squares and put the roots of each factor in the proper place. $(6y^2 + 2z)(6y^2 - 2z)$
Check using FOIL. $\underline{(6y^2 + 2z)(6y^2 - 2z) = 36y^4 - 12y^2z + 12y^2z - 4z^2 = 36y^4 - 4z^2}$

370. This expression can be factored using the trinomial method. The factors of c^2 are c and c, and the factors of 32 are (1)(32) or (2)(16) or (4)(8). The sign of the numerical term is positive, so the signs in the factors of our trinomial factorization must be the same. The sign of the first-degree term (the variable to the power of 1) is negative. This leads one to believe that the signs in the trinomial factors will both be negative. The only factors of 32 that add up to 12 are 4 and 8.
Check using FOIL. $\underline{(c - 4)(c - 8) = c^2 - 8c - 4c + 32 = c^2 - 12c + 32}$

371. First, the terms have a common factor of d. Factor d out of each term and write the expression in factored form. $d(d^2 - 16d + 55)$

The expression within parentheses can be factored using the trinomial method. The factors of d^2 are d and d, and the factors of 55 are (1)(55) and (5)(11). You want the result of the O and I of the FOIL method for multiplying factors to add up to ^-16d. Only terms with the same sign will result in a positive numerical term, which is what you need, since the third term is 55. Place the factors (5)(11) into the trinomial factor form and check using FOIL. $d^2 - 16d + 55 = (d - 5)(d - 11) = d^2 - 11d - 5d + 55$

Substitute the factored form of the above expression into the expression obtained in the first step to get the completely factored form of the original trinomial. $\underline{d^3 - 16d^2 + 55d = d(d - 5)(d - 11)}$

372. First, the terms have a common factor of w^2. Factor w^2 out of each term and write the expression in factored form. $w^4 + 17w^3 + 60w^2 = w^2(w^2 + 17w + 60)$

The expression within parentheses can be factored using the trinomial method. The factors of w^2 are w and w, and the factors of 60 are (1)(60), (2)(30), (3)(20), (4)(15), (5)(12), and (6)(10). You want the result of the O and I of the FOIL method for multiplying factors to add up to $17w$. Only terms with the same sign will result in a positive numerical term, which is what you need, since the third term is 60. Place the factors (5)(12) into the trinomial factor form and check using FOIL. $w^2 + 17w + 60 = (w + 5)(w + 12) = w^2 + 12w + 5w + 60$

Substitute the factored form of the above expression into the expression obtained in the first step to get the completely factored form of the original trinomial. $\underline{w^4 + 17w^3 + 60w^2 = w^2(w + 5)(w + 12)}$

373. First, the terms have a common factor of y^3. Factor y^3 out of each term and write the expression in factored form. $y^5 + 16y^4 + 28y^3 = y^3(y^2 + 16y + 28)$

This expression can be factored using the trinomial method. The factors of y^2 are y and y, and the factors of 28 are (1)(28), (2)(14), and (4)(7). You want the result of the O and I of the FOIL method for multiplying factors to add up to $16y$. Only terms with the same sign will result in a positive numerical term, which is what you need, since the third term is 28. Place the factors (2)(14) into the trinomial factor form and check using FOIL. $y^2 + 16y + 28 = (y + 2)(y + 14) = y^2 + 14y + 2y + 28$

Substitute the factored form of the above expression into the expression obtained in the first step to get the completely factored form of the original trinomial. $\underline{y^5 + 16y^4 + 28y^3 = y^3(y + 2)(y + 14)}$

374. This expression can be factored using the trinomial method. The factors of v^4 are $(v^2)(v^2)$, and the factors of 48 are (1)(48) or (2)(24) or (3)(16) or (4)(12) or (6)(8). Only the product of a positive and a negative numerical term will result in $^-48$. The only factors of 48 that can be added or subtracted in any way to equal 13 are 3 and 16. Use 3 and 16 and a positive and negative sign in the terms of the trinomial factors. Check your answer using FOIL. $\underline{(v^2 + 3)(v^2 - 16) = v^4 - 16v^2 + 3v^2 - 48 = v^4 - 13v^2 - 48}$

You may notice that one of the two factors of the trinomial expression can itself be factored. The second term is the difference of two perfect squares. Factor $(v^2 - 16)$ using the form for factoring the difference of two perfect squares. $(v + 4)(v - 4) = v^2 - 4v + 4v - 16 = v^2 - 16$

This now makes the complete factorization of $\underline{v^4 - 13v^2 - 48 = (v^2 + 3)(v + 4)(v - 4).}$

375. The factors of x^2 are x and x, and the factors of 36 are (1)(36) or (2)(18) or (4)(9) or (6)(6). The sign of the numerical term is positive, so the signs in the factors of our trinomial factorization must be the same. The sign of the first-degree term (the variable to the power of 1) is negative. This leads one to believe that the signs in the trinomial factors will both be negative. The only factors of 36 that add up to 20 are 2 and 18. Use them and two negative signs in the trinomial factor form.

Check your answer using FOIL. $\underline{(x - 2)(x - 18) = x^2 - 18x - 2x + 36 = x^2 - 20x + 36}$

16

Using Factoring

This chapter will present polynomial expressions for you to factor. In the previous chapter, all the coefficients of the second-degree terms were 1. In this chapter, the coefficients of the second-degree terms will often be whole numbers greater than 1. This will complicate the process of factoring by adding more possibilities to check. In some cases, you will find that you can factor an expression using more than one of the three methods of factoring polynomials.

Always look to factor algebraic expressions using the greatest common factor method first. Then analyze the remaining expression to determine if other factoring methods can be used. The three methods for factoring polynomial expressions are:

1. Greatest common factor method
2. Difference of two perfect squares method
3. Trinomial method

When presented with a polynomial with a coefficient greater than 1 for the second-degree term, use the trinomial factor form $(ax \pm (\ ))(bx \pm (\ ))$ where $a \cdot b =$ the coefficient of the second-degree term. List the factors of the numerical term of the trinomial and consider the choices of factors and signs that will result in the correct trinomial factorization.

After choosing terms to try in the trinomial factors form, use FOIL to check your guesses for the trinomial factors. You will want to do a partial check by first completing the O and the I part of FOIL to determine if you have the first-degree term right.

Tips for Using Factoring

When factoring a trinomial expression, first determine the signs that will be used in the two factors.

Next, list the possible factors of the second-degree term.

Then list the factors of the numerical term.

Finally, place the factors into the trinomial factor form in all possible ways and use FOIL to check for the correct factorization.

Be systematic in your attempts to be sure you try all possible choices. You will become better at factoring as you learn to look for the combinations of factors that will give you the required results for the first-degree term.

Factor the following expressions.

376. $2x^2 + 7x + 6$

377. $3x^2 + 13x + 12$

378. $5x^2 - 14x - 3$

379. $9x^2 + 15x + 4$

380. $10x^2 + 33x - 7$

381. $3x^2 - 3x - 18$

382. $4a^2 - 16a - 9$

383. $8x^2 + 10x + 3$

384. $6a^2 - 5a - 6$

385. $16y^2 - 100$

386. $6x^2 + 15x - 36$

387. $4bc^2 + 22bc - 42b$

388. $2a^6 + a^3 - 21$

389. $6a^2x - 39ax - 72x$

390. $36m^2 - 102m + 30$

391. $5c^2 - 9c - 2$

392. $9x^3 - 4x$

393. $6y^3 + 5y^2 + y$

394. $4x^4 - 37x^2 + 9$

395. $12z^4 + 13z^2 + 3$

396. $4xy^3 + 6xy^2 - 10xy$

397. $4ax^2 - 38ax - 66a$

398. $3c^2 + 19c - 40$

399. $^-15d^4 + 5d^3 + 10d^2$

400. $4x^4 + 2x^2 - 30$

Answers

Numerical expressions in parentheses like this [ ] are operations performed on only part of the original expression. The operations performed within these symbols are intended to show how to evaluate the various terms that make up the entire expression.

Expressions with parentheses that look like this ( ) contain either numerical substitutions or expressions that are part of a numerical expression. Once a single number appears within these parentheses, the parentheses are no longer needed and need not be used the next time the entire expression is written.

When two pair of parentheses appear side by side like this ( )( ), it means that the expressions within are to be multiplied.

Sometimes parentheses appear within other parentheses in numerical or algebraic expressions. Regardless of what symbol is used, ( ), { }, or [ ], perform operations in the innermost parentheses first and work outward.

<u>Underlined</u> expressions show the original algebraic expression as an equation with the expression equal to its factored result.

376. Both signs in the trinomial are positive, so use positive signs in the trinomial factor form.

$(ax + (\ ))(bx + (\ ))$

The factors of the second-degree term are $2x^2 = (2x)(x)$.

The numerical term $6 = (1)(6) = (2)(3)$.

You want to get $7x$ from adding the result of the Outer and Inner multiplications when using FOIL. You could make the following guesses for the factors of the original expression.

$(2x + (1))(x + (6))$

$(2x + (6))(x + (1))$

$(2x + (2))(x + (3))$

$(2x + (3))(x + (2))$

Now just consider the results of the Outer and Inner products of the terms for each guess. The one that results in a first-degree term of $7x$ is the factorization you want to fully check.

$(2x + (1))(x + (6))$ will result in Outer product plus Inner product:
$2x(6) + (1)x = 12x + x = 13x$.

$(2x + (6))(x + (1))$ will result in Outer product plus Inner product:
$2x(1) + (6)x = 2x + 6x = 8x$.

$(2x + (2))(x + (3))$ will result in Outer product plus Inner product:
$2x(3) + (2)x = 6x + 2x = 8x$.

$(2x + (3))(x + (2))$ will result in Outer product plus Inner product:
$2x(2) + (3)x = 4x + 3x = 7x$.

Place the factors in the trinomial factor form so that the product of the outer terms $(2x)(2) = 4x$ and the product of the inner terms $(3)(x) = 3x$. That way, $4x + 3x = 7x$, the middle term of the trinomial.

$(2x + (3))(x + (2))$

Check using FOIL.

First—$(2x)(x) = 2x^2$

Outer—$(2x)(2) = 4x$
Inner—$(3)(x) = 3x$
Last—$(3)(2) = 6$
Add the products of multiplication using FOIL. $2x^2 + 4x + 3x + 6 = 2x^2 + 7x + 6$
The factors check out. $\underline{(2x + 3)(x + 2) = 2x^2 + 7x + 6}$

377. Both signs in the trinomial are positive, so use positive signs in the trinomial factor form.
$(ax + (\ ))(bx + (\ ))$
The factors of the second-degree term are $3x^2 = (3x)(x)$.
The factors of the numerical term $12 = (1)(12) = (2)(6) = (3)(4)$.
You want to get $13x$ from adding the result of the Outer and Inner multiplications when using FOIL. Place the factors in the trinomial factor form so that the product of the outer terms $(3x)(3) = 9x$ and the product of the inner terms $(4)(x) = 4x$. Then $9x + 4x = 13x$, the middle term of the trinomial.
$(3x + 4)(x + 3)$
Check using FOIL.
First—$(3x)(x) = 3x^2$
Outer—$(3x)(3) = 9x$
Inner—$(4)(x) = 4x$
Last—$(4)(3) = 12$
Add the products of multiplication using FOIL.
$3x^2 + 9x + 4x + 12 = 3x^2 + 13x + 12$
The factors check out. $\underline{(3x + 4)(x + 3) = 3x^2 + 13x + 12}$

378. Both signs in the trinomial are negative. To get a negative sign for the numerical term, the signs in the factors must be + and -.
$(ax + (\ ))(bx - (\ ))$
The factors of the second-degree term are $5x^2 = (5x)(x)$.
The factors of the numerical term $3 = (1)(3)$.
When you multiply the Outer and Inner terms of the trinomial factors, the results must add up to be ^-14x. Multiplying, $5x(^-3) = {}^-15x$, and $1x(^+1) = {}^+1x$. Adding $(^-15x) + (^+1x) = {}^-14x$. Place those terms into the trinomial factor form.
$(5x + (1))(x - (3))$
Check using FOIL.
First—$(5x)(x) = 5x^2$
Outer—$(5x)(^-3) = {}^-15x$
Inner—$(1)(x) = x$
Last—$(1)(^-3) = {}^-3$
Add the products of multiplication using FOIL.
$5x^2 - 15x + 1x - 3 = 5x^2 - 14x - 3$
The factors check out. $\underline{(5x + 1)(x - 3) = 5x^2 - 14x - 3}$

379. Both signs in the trinomial are positive, so use positive signs in the trinomial factor form. $(ax + (\ ))(bx + (\ ))$

The factors of the second-degree term $9x^2 = (9x)(x)$ or $9x^2 = (3x)(3x)$.

The factors of the numerical term $4 = (1)(4)$ or $4 = (2)(2)$

To get $15x$ from adding the result of the Outer and Inner multiplications when using FOIL, place the factors in the trinomial factor form so that the product of the outer terms $(3x)(1) = 3x$ and the product of the inner terms $(4)(3x) = 12x$. Then $3x + 12x = 15x$, the middle term of the trinomial. $(3x + 4)(3x + 1)$

Check using FOIL.

First—$(3x)(3x) = 9x^2$

Outer—$(3x)(1) = 3x$

Inner—$(4)(3x) = 12x$

Last—$(4)(1) = 4$

The factors check out. $(3x + 4)(3x + 1) = 9x^2 + 3x + 12x + 4 = 9x^2 + 15x + 4$

380. The sign of the numerical term is negative. So, the signs in the trinomial factor form will have to be + and –. $(ax + (\ ))(bx - (\ ))$.

The factors of the second-degree term are $10x^2 = (10x)(x)$ or $10x^2 = (2x)(5x)$.

The factors of the numerical term 7 are $(1)(7)$.

When you multiply the Outer and Inner terms of the trinomial factors, the results must add up to 33. Multiplying, $(5x)(7) = 35x$ and $(2x)(-1) = {}^-2x$. Adding, $35x - 2x = 33x$.

Place those terms into the trinomial factor form: $(5x - 1)(2x + 7)$

Check using FOIL.

First—$(5x)(2x) = 10x^2$

Outer—$(5x)(7) = 35x$

Inner—$({}^-1)(2x) = {}^-2x$

Last—$({}^-1)(7) = {}^-7$

Add the products of the multiplication using FOIL: $10x^2 + 35x - 2x - 7$

The factors check out. $10x^2 + 33x - 7 = (5x - 1)(2x + 7)$

381. The three terms have a common factor of 3. You can factor out 3 and represent the trinomial as $3(x^2 - x - 6)$. Now factor the trinomial in the parentheses, and don't forget to include the factor 3 when you are done.

The sign of the numerical term is negative. So the signs in the trinomial factor form will have to be + and –, because that is the only way to get a negative sign when multiplying the Last terms when checking with FOIL. $(ax + (\ ))(bx - (\ ))$

The factors of the second-degree term are $3x^2 = (3x)(x)$.

The sign of the second term is negative. That tells you that the result of adding the products of the Outer and Inner terms of the trinomial factors must result in a negative sum for the x term. The factors of the numerical term 6 are $(1)(6)$ or $(2)(3)$. Put the + with the 2 and the – with the 3. $(x + 2)(x - 3)$

Check using FOIL.
First—$(x)(x) = x^2$
Outer—$(x)(^-3) = {}^-3x$
Inner—$(2)(x) = 2x$
Last—$(2)(^-3) = {}^-6$
The factors check out. $(x + 2)(x - 3) = x^2 - 3x + 2x - 6 = x^2 - x - 6$
Include the common factor of 3 so that
$3(x + 2)(x - 3) = 3(x^2 - x - 6) = 3x^2 - 3x - 18.$

382. Both signs in the trinomial expression are negative. To get a negative sign for the numerical term, the signs within the trinomial factors must be + and –. $(ax + (\ ))(bx - (\ ))$
The factors of the second-degree term $4a^2 = (4a)(a)$ or $4a^2 = (2a)(2a)$.
The factors of the numerical term 9 are (1)(9) or (3)(3).
The coefficient of the first-degree term is 2 less than 18. You can multiply $2a$ and (9) to get $18a$ leaving the factors $2a$ and (1) to get a $2a$. Use this information to place factors within the trinomial factor form.
$(2a + 1)(2a - 9)$
Check using FOIL.
First—$(2a)(2a) = 4a^2$
Outer—$(2a)(^-9) = {}^-18a$
Inner—$(1)(2a) = 2a$
Last—$(1)(^-9) = {}^-9$
The result of multiplying the factors is
$(2a + 1)(2a - 9) = 4a^2 - 18a + 2a - 9 = 4a^2 - 16a - 9.$

383. Both signs in the trinomial expression are positive. So, the signs in the trinomial factor form will have to be + and +. $(ax + (\ ))(bx + (\ ))$.
The factors of the second-degree term are $8x^2 = (8x)(x)$ or $8x^2 = (2x)(4x)$.
The factors of the numerical term 3 are (1)(3).
When you multiply the Outer and Inner terms of the trinomial factors, the results must add up to $10x$. Multiplying,
$(4x)(1) = 4x$ and $(2x)(3) = 6x$. Adding, $4x + 6x = 10x$.
Place those terms into the trinomial factor form: $(4x + 3)(2x + 1)$
Check using FOIL.
First—$(4x)(2x) = 8x^2$
Outer—$(4x)(1) = 4x$
Inner—$(3)(2x) = 6x$
Last—$(3)(1) = 3$
Add the products of the multiplication using FOIL: $8x^2 + 4x + 6x + 3$

The factors check out. $8x^2 + 10x + 3 = (4x + 3)(2x + 1)$

384. Both signs in the trinomial expression are negative. To get a negative sign for the numerical term, the signs within the trinomial factors must be + and –. $(ax + (\ ))(bx - (\ ))$

The factors of the second-degree term $6a^2 = (6a)(a)$ or $6a^2 = (2a)(3a)$.

The factors of the numerical term 6 are (6)(1) or (2)(3).

The trinomial looks balanced with a 6 on each end and a 5 in the middle. Try a balanced factor arrangement. $(3a + 2)(2a - 3)$

Check using FOIL.

First—$(3a)(2a) = 6a^2$

Outer—$(3a)(^-3) = {}^-9a$

Inner—$2(2a) = 4a$

Last—$(2)(^-3) = {}^-6$

Combining the results of multiplying using FOIL results in $\underline{(3a + 2)(2a - 3) = 6a^2 - 9a + 4a - 6 = 6a^2 - 5a - 6}$.

385. This expression is the difference between two perfect squares. Using the form for the difference of two perfect squares gives you the factors $(4y + 10)(4y - 10)$.

However, there is a greatest common factor that could be factored out first to leave $4(4y^2 - 25)$. Now you need only factor the difference of two simpler perfect squares.

$4(4y^2 - 25) = 4(2y + 5)(2y - 5)$

The first factorization is equivalent to the second because you can factor out two from each of the factors. This will result in $\underline{(2)(2)(2y + 5)(2y - 5) \text{ or } 4(2y + 5)(2y - 5)}$. When factoring polynomials, watch for the greatest common factors first.

386. The terms of the trinomial have a greatest common factor of 3. So the term $3(2x^2 + 5x - 12)$ will simplify the trinomial factoring. You need only factor the trinomial within the parentheses.

The sign of the numerical term is negative. So the signs in the trinomial factor form will have to be + and –. $(ax + (\ ))(bx - (\ ))$

The factors of the second-degree term are $2x^2 = (2x)(x)$.

The factors of the numerical term 12 are (1)(12) or (2)(6) or (3)(4).

The factors $(2x)(4) = 8x$ and the remaining factors $(x)(3) = 3x$. It's clear that $8x - 3x = 5x$. Use those factors in the trinomial factor form.

$(2x - 3)(x + 4)$

Check using FOIL.

First—$(2x)(x) = 2x^2$

Outer—$(2x)(4) = 8x$

Inner—$(^-3)(x) = {}^-3x$

Last—$(^-3)(4) = {}^-12$

The result of multiplying factors is $2x^2 + 8x - 3x - 12 = 2x^2 + 5x - 12$.

Now include the greatest common factor of 3 for the final solution.

$\underline{3(2x - 3)(x + 4) = 6x^2 + 15x - 36}$

387. Each term in the polynomial has a common factor of $2b$. The resulting expression looks like this: $2b(2c^2 + 11c - 21)$

The sign of the numerical term is negative. So the signs in the trinomial factor form will have to be + and – because that is the only way to get a negative sign when multiplying the Last terms when checking with FOIL. $(ax + ( ))(bx - ( ))$

The factors of the second-degree term $2c^2 = (2c)(c)$.

The factors of the numerical term 21 are (1)(21) or (3)(7). The factors $(2c)(7) = 14c$ and the associated factors $(c)(3) = 3c$. Place these factors in the trinomial factor form so that the result of the Outer and Inner products when using FOIL to multiply are ^+14c and ^-3c. $(2c - 3)(c + 7)$

Check using FOIL.

First—$(2c)(c) = 2c^2$

Outer—$(2c)(7) = 14c$

Inner—$(^-3)(c) = {}^-3c$

Last—$(^-3)(7) = {}^-21$

The product of the factors is

$(2c - 3)(c + 7) = 2c^2 + 14c - 3c - 21 = 2c^2 + 11c - 21$.

Now include the greatest common factor term.

$2b(2c - 3)(c + 7) = 2b(2c^2 + 11c - 21) = 4bc^2 + 22bc - 42b$

388. This expression appears to be in the familiar trinomial form, but what's with those exponents? Think of $a^6 = (a^3)^2$. Then the expression becomes $2(a^3)^2 + (a^3) - 21$. Now you factor like it was a trinomial expression. The sign of the numerical term is negative. So the signs in the trinomial factor form will have to be + and –. $(ax + ( ))(bx - ( ))$

The second-degree term $2(a^3)^2 = 2(a^3)(a^3)$.

The factors of the numerical term 21 are (1)(21) or (3)(7). The factors $(a^3)(7) = 7(a^3)$ and the factors $(2(a^3))(3) = 6(a^3)$. The difference of 7 and 6 is 1. Place these factors in the trinomial factor form so that the first degree term is $1(a^3)$.

$(a^3 - 3)(2a^3 + 7)$

Check using FOIL.

First—$(a^3)(2a^3) = 2a^6$

Outer—$(a^3)(7) = 7a^3$

Inner—$(^-3)(2a^3) = {}^-6a^3$

Last—$(^-3)(7) = {}^-21$

The product of the factors is

$(a^3 - 3)(2a^3 + 7) = 2a^6 + 7a^3 - 6a^3 - 21 = 2a^6 + a^3 - 21$.

The factors of the trinomial are correct.

389. The greatest common factor of the terms in the trinomial expression is $3x$. Factoring $3x$ out results in the expression $3x(2a^2 - 13a - 24)$. Factor the trinomial expression inside the parentheses.

The sign of the numerical term is negative. So the signs in the trinomial factor form will have to be + and –. $(ax + (\ ))(bx - (\ ))$

The factors of the term $2a^2 = (2a)(a)$.

The factors of the numerical term 24 are (1)(24) or (2)(12) or (3)(8) or (4)(6).

The factors $(2a)(8) = 16a$, and the related factors $(a)(3) = 3a$. The difference of 16 and 3 is 13. Place these numbers in the trinomial factor form, and check the expression using FOIL.

$(2a + 3)(a - 8)$

First—$(2a)(a) = 2a^2$

Outer—$(2a)(^-8) = {}^-16a$

Inner—$(3)(a) = 3a$

Last—$(3)(^-8) = {}^-24$

The result is $(2a + 3)(a - 8) = 2a^2 - 16a + 3a - 24 = 2a^2 - 13a - 24$.

Now include the greatest common factor if $3x$.

$\underline{3x(2a + 3)(a - 8) = 3x(2a^2 - 13a - 24) = 6a^2x - 39ax - 72x}$

390. The three terms have a common factor of 6. You can factor out 6 and represent the trinomial as $6(6m^2 - 17m + 5)$. Now, factor the trinomial in parentheses, and don't forget to include the factor 6 when you are done.

The sign of the numerical term is positive, and the sign of the middle term is negative. So, the signs in the trinomial factor form will have to be – and –. $(am - (\ ))(bm - (\ ))$.

The factors of the second-degree term are $6m^2 = (6m)(m)$ or $6m^2 = (2m)(3m)$.

The factors of the numerical term 5 are (1)(5).

When you multiply the Outer and Inner terms of the trinomial factors, the results must add up to $^-17$. Multiplying, $(3m)(^-5) = {}^-15m$ and $(^-1)(2m) = {}^-2m$. Adding, $^-15m - 2m = {}^-17m$.

Place those terms into the trinomial factor form: $(3m - 1)(2m - 5)$

Check using FOIL.

First—$(3)(2) = 6m2$

Outer—$(3m)(-5) = {}^-15m$

Inner—$(^-1)(2m) = {}^-2m$

Last—$(^-1)(^-5) = 5$

Add the products of the multiplication using FOIL: $6m^2 - 15m - 2m + 5$

The factors check out. $\underline{36m^2 - 102m + 30 = 6(3m - 1)(2m - 5)}$

391. The numerical term of the trinomial has a negative sign, so the signs within the factors of the trinomial will be a + and –. $(ax + (\ ))(bx - (\ ))$
The only factors of the second-degree term are $(c)(5c)$.
The numerical term of the trinomial 2 has factors of (1)(2). What combination will result in a ^-9c when the Outer and Inner products of the trinomial factors are added together?
Our choices are $(5)(1)c + (1)(^-2)c$, $(5)(^-1)c + (1)(2)c$, $(5)(2)c + (1)(^-1)c$, $(5)(^-2)c + (1)(1)c$. The last of these is equal to the desired ^-9c, which gives the factoring $(5c + 1)(c - 2)$. Check using FOIL to multiply terms. $(5c + 1)(c - 2)$
Check using FOIL.
First—$(5c)(c) = 5c^2$
Outer—$(5c)(^-2) = {^-10c}$
Inner—$(1)(c) = c$
Last—$(1)(^-2) = {^-2}$
The product of the factors $\underline{(5c + 1)(c - 2) = 5c^2 - 10c + c - 2 = 5c^2 - 9c - 2}$.

392. The terms of the expression have a greatest common factor of x. Factoring x out of the expression results in $x(9x^2 - 4)$.
The expression inside the parentheses is the difference of two perfect squares. Factor that expression using the form for the difference of two perfect squares. Include the greatest common factor to complete the factorization of the original expression.
$9x^2 = (3x)^2$
$4 = 2^2$
Using the form, the factorization of the difference of two perfect squares is $(3x - 2)(3x + 2)$.
Check using FOIL.
First—$(3x)(3x) = 9x^2$
Outer—$(3x)(2) = 6x$
Inner—$(^-2)(3x) = {^-6x}$
Last—$(^-2)(2) = {^-4}$
Include the greatest common factor x in the complete factorization.
$\underline{x(3x - 2)(3x + 2) = x(9x^2 - 4) = 9x^3 - 4x}$

393. The three terms have a common factor of y. You can factor out y and represent the trinomial as $y(6y^2 + 5y + 1)$. Now, factor the trinomial in parentheses, and don't forget to include the factor y when you are done.
Both signs in the trinomial expression are positive. So, the signs in the trinomial factor form will have to be + and +. $(ay + (\ ))(by + (\ ))$.
The factors of the second-degree term are $6y^2 = (6y)(y)$ or $6y^2 = (2y)(3y)$.
The factors of the numerical term 1 are (1)(1).
When you multiply the Outer and Inner terms of the trinomial factors, the results must add up to $5y$. Multiplying,
$(2y)(1) = 2y$ and $(1)(3y) = 3y$. Adding, $2y + 3y = 5y$.
Place those terms into the trinomial factor form: $(2y + 1)(3y + 1)$

Check using FOIL.
First—$(2y)(3y) = 6y^2$
Outer—$(2y)(1) = 2y$
Inner—$(1)(3y) = 3y$
Last—$(1)(1) = 1$
Add the products of the multiplication using FOIL: $6y^2 + 2y + 3y + 1$
The factors check out. $\underline{6y^3 + 5y^2 + y = y(2y + 1)(3y + 1)}$

394. When you think of $x^4 = (x^2)^2$, you can see that the expression is a trinomial that is easy to factor.
The numerical term is positive, so the signs in the trinomial factor form will be the same. The sign of the first-degree term is negative, so you will use two – signs. $(ax - (\ ))(bx - (\ ))$
The factors of the second-degree term are $4x^4 = (x^2)(4x^2)$ or $(2x^2)(2x^2)$.
The numerical term 9 has (1)(9) or (3)(3) as factors.
What combination will result in a total of 37 when the Outer and Inner products are determined? $4x^2(9) = 36x^2$, $1x^2\ (1) = 1x^2$ and $36x^2 + 1x^2 = 37x^2$. Use these factors in the trinomial factor form.
$(4x^2 - 1)(x^2 - 9)$
Check using FOIL and you will find
$(4x^2 - 1)(x^2 - 9) = 4x^4 - 36x^2 - x^2 + 9 = 4x^4 - 37x^2 + 9$.
Now you need to notice that the factors of the original trinomial expression are themselves both factorable. Why? Because they are both the difference of two perfect squares.
Use the factor form for the difference of two perfect squares for each factor of the trinomial.
$(4x^2 - 1) = (2x + 1)(2x - 1)$
$(x^2 - 9) = (x + 3)(x - 3)$
Put the factors together to complete the factorization of the original expression.
$\underline{(4x^2 - 1)(x^2 - 9) = (2x + 1)(2x - 1)(x + 3)(x - 3)}$

395. This expression is in trinomial form. If you think of the variable as z^2, you can see that the expression is in trinomial form. Use z^2 where you usually put a first-degree variable. The trinomial you will be factoring looks like this: $12(z^2)^2 + 13(z^2) + 3$
Both signs in the trinomial expression are positive. So, the signs in the trinomial factor form will have to be + and + . $\left(a(z^2) + (\ )\right)\left(b(z^2) + (\ )\right)$.
The factors of the second-degree term are $12(z^2)^2 = [12(z^2)][(z^2)]$, or $12(z^2)^2 = [2(z^2)][6(z^2)]$, or $12(z^2)^2 = [3(z^2)][4(z^2)]$.
The factors of the numerical term 3 are (1)(3).
When you multiply the Outer and Inner terms of the trinomial factors, the results must add up to $13(z^2)$. Multiplying, $[4(z^2)](1) = 4(z^2)$ and $(3)[3(z^2)] = 9(z^2)$. Adding, $4(z^2) + 9(z^2) = 13(z^2)$.

Place those terms into the trinomial factor form: $(4z^2 + 3)(3z^2 + 1)$
Check using FOIL.
First—$(4z^2)(3z^2) = 12z^4$
Outer—$(4z^2)(1) = 4z^2$
Inner—$(3)(3z^2) = 9z^2$
Last—$(3)(1) = 3$
Add the products of the multiplication using FOIL: $12z^4 + 4z^2 + 9z^2 + 3$
The factors check out. $\underline{12z^4 + 13z^2 + 3 = (4z^2 + 3)(3z^2 + 1)}$

396. Each term in the expression has a common factor of $2xy$. When factored out, the expression becomes $2xy(2y^2 + 3y - 5)$. Now factor the trinomial in the parentheses.
The last sign is negative, so the signs within the factor form will be a + and –.
$(ax + (\ ))(bx - (\ ))$
The factors of the second-degree term are $2y^2 = y(2y)$.
The numerical term 5 has factors (5)(1).
Place the factors of the second degree and the numerical terms so that the result of the Outer and Inner multiplication of terms within the factor form of a trinomial expression results in a ^+3x.
$(2y + 5)(y - 1)$
Multiply using FOIL. $(2y + 5)(y - 1) = 2y^2 - 2y + 5y - 5 = 2y^2 + 3y - 5$
The factors of the trinomial expression are correct. Now include the greatest common factor to complete the factorization of the original expression.
$\underline{2xy(2y + 5)(y - 1) = 2xy(2y^2 - 2y + 5y - 5) = 2xy(2y^2 + 3y - 5)}$

397. The terms of the trinomial have a greatest common factor of $2a$. When factored out, the resulting expression is $2a(2x^2 - 19x - 33)$. The expression within the parentheses is a trinomial and can be factored.
The signs within the terms of the factor form will be + and – because the numerical term has a negative sign. Only a $(+)(-) = (-)$. $(ax + (\ ))(bx - (\ ))$
The factors of the second-degree term are $2^2 = (2)$.
The numerical term 33 has (1)(33) or (3)(11) as factors.
Since $2a(11) = 22a$, and $a(3) = 3a$, and $22a - 3a = 19a$, use those factors in the trinomial factor form so that the result of the multiplication of the Outer and Inner terms results in ^-19x.
$(2x + 3)(x - 11)$
Check using FOIL. $(2x + 3)(x - 11) = 2x^2 - 22x + 3x - 33 = 2x^2 - 19x - 33$
The factorization of the trinomial factor is correct. Now include the greatest common factor of the original expression to get the complete factorization of the original expression.
$\underline{2a(2x + 3)(x - 11) = 2a(2x^2 - 22x + 3x - 33) = 2a(2x^2 - 19x - 33)}$

398. The signs within the terms of the factor form will be + and – because the numerical term has a negative sign. $(ax + (\ ))(bx - (\ ))$
The factors of the second-degree term are $3c^2 = c(3c)$.
The numerical term 40 has (1)(40) or (2)(20) or (4)(10) or (5)(8) as factors.
You want the result of multiplying and then adding the Outer and Inner terms of the trinomial factor form to result in a $^{+}19c$ when the like terms are combined. Using trial and error, you can determine that $3c(8) = 24c$, and $c(5) = 5c$, and $24c - 5c = 19c$. Use those factors in the factor form in such a way that you get the result you seek.
$(3c - 5)(c + 8) = 3c^2 + 24c - 5c - 40 = 3c^2 + 19c - 40$
The complete factorization of the original expression is $\underline{(3c - 5)(c + 8)}$.

399. The three terms have a common factor of $^{-}5d^2$. You can factor out $^{-}5d^2$ and represent the trinomial as $^{-}5d^2(3d^2 - d - 2)$. Now, factor the trinomial in parentheses, and don't forget to include the factor $^{-}5d^2$ when you are done. (NOTE! We pulled out $^{-}1$ as part of the common factor to ensure that the coefficient of the second degree term was positive.)
Both signs in the trinomial expression are negative. So, the signs in the trinomial factor form will have to be - and +. $(ad - (\ ))(bd + (\ ))$.
The factors of the second-degree term are $3d^2 = (3d)(d)$.
The factors of the numerical term 2 are (1)(2).
When you multiply the Outer and Inner terms of the trinomial factors, the results must add up to -. Multiplying,
$(3d)(^{-}1) = {}^{-}3d$ and $(d)(2) = 2$. Adding, $^{-}3d + 2d = {}^{-}d$.
Place those terms into the trinomial factor form: $(d - 1)(3d + 2)$
Check using FOIL.
First—$(d)(3d) = 3d^2$
Outer—$(d)(2) = 2d$
Inner—$(^{-}1)(3d) = {}^{-}3d$
Last—$(^{-}1)(2) = {}^{-}2$
Add the products of the multiplication using FOIL: $3d^2 + 2d - 3d - 2$
The factors check out. $\underline{^{-}15d^4 + 5d^3 + 10d^2 = {}^{-}5d^2(d - 1)(3d + 2)}$

400. This expression is in trinomial form. If you think of the variable as x^2, you can see that the expression is in the trinomial form. Use x^2 where you usually put a first-degree variable. The trinomial you will be factoring looks like this: $4(x^2)^2 + 2(x^2) - 30$.
The signs within the terms of the factor form will be + and – because the numerical term has a negative sign.
$(ax^2 + (\ ))(bx^2 - (\ ))$
The term $(2x^2)^2$ can be factored as $(2x^2)^2 = x^2(4x^2)$ or $(2x^2)(2x^2)$.
The numerical term 30 can be factored as (1)(30) or (2)(15) or (3)(10) or (5)(6).

The factors $(4x^2)(3) = 12x^2$ and $x^2(10) = 10x^2$ will give you $12x^2 - 10x^2 = 2x^2$ when you perform the Inner and Outer multiplications and combine like terms using FOIL with the terms in the trinomial factor form. The factors of the expression will be $(4x^2 - 10)(x^2 + 3)$.

Check using FOIL.

$$(4x^2 - 10)(x^2 + 3) = 4x^4 + 12x^2 - 10x^2 - 30 = 4x^4 + 2x^2 - 30$$

The expression $(4x^2 - 10)(x^2 + 3)$ is the correct factorization of the original expression. However, the first factor has a greatest common factor of 2. So a complete factorization would be $2(2x^2 - 5)(x^2 + 3)$.

Did you notice that you could have used the greatest common factor method to factor out a 2 from each term in the original polynomial? If you did, you would have had to factor the trinomial expression $2x^4 + x^2 - 15$ and multiply the result by the factor 2 to equal the original expression. Let's see:

$$\underline{2(2x^4 + x^2 - 15) = 2(2x^2 - 5)(x^2 + 3) = (4x^2 - 10)(x^2 + 3) = 4x^4 + 2x^2 - 30}$$

It all comes out the same, but if you left the factor of 2 in the term $(4x^2 - 10)$, the end result would not be a complete factorization of the original trinomial expression.

17

Solving Quadratic Equations

This chapter will give you practice in finding solutions to quadratic equations. Quadratic equations are those equations that can be written in the form $ax^2 + bx + c = 0$, where $a \neq 0$. While there are several methods for solving quadratic equations, solutions for all the equations presented here can be found by factoring.

In the previous two chapters, you practiced factoring polynomials by using the greatest common factor method, the difference of two perfect squares method, and the trinomial factor method. Use these methods to factor the equations that have been transformed into quadratic equations. Then, using the zero product property (if $(a)(b) = 0$, then $a = 0$ or $b = 0$ or both $= 0$), let each factor equal zero and solve for the variable. There will be two solutions for each quadratic equation. (Ignore numerical factors such as the 3 in the factored equation $3(x + 1)(x + 1) = 0$ when finding solutions to quadratic equations. The solutions will be the same for equations with or without the numerical factors.)

Find the solutions to the following quadratic equations.

401. $x^2 - 25 = 0$

402. $n^2 - 169 = 0$

403. $a^2 + 12a + 32 = 0$

404. $y^2 - 15y + 56 = 0$

405. $b^2 + b - 90 = 0$

406. $4x^2 = 49$

407. $81z^2 = 324$

408. $2n^2 + 20n + 42 = 0$

409. $3c^2 - 33c - 78 = 0$

410. $6x^2 = 17x - 5$

411. $3x^2 - 36x + 108 = 0$

412. $7a^2 - 21a - 28 = 0$

413. $8y^2 + 56y + 96 = 0$

414. $2x^2 + 9x = {}^-10$

415. $4x^2 + 4x = 15$

416. $20 = 35x - 15x^2$

417. $3x^2 = 19x - 20$

418. $8b^2 + 10b = 42$

419. $17x - 30x^2 = 1$

420. $6b^2 + 20b = {}^-9b - 20$

421. $15x^2 - 70x - 120 = 0$

422. $7x^2 = 52x - 21$

423. $36z^2 + 78z = {}^-36$

424. $200z^2 - 10z - 3 = 0$

425. $24x^2 = 3(43x - 15)$

Answers

Numerical expressions in parentheses like this [ ] are operations performed on only part of the original expression. The operations performed within these symbols are intended to show how to evaluate the various terms that make up the entire expression.

Expressions with parentheses that look like this ( ) contain either numerical substitutions or expressions that are part of a numerical expression. Once a single number appears within these parentheses, the parentheses are no longer needed and need not be used the next time the entire expression is written.

When two pair of parentheses appear side by side like this ( )( ), it means that the expressions within are to be multiplied.

Sometimes parentheses appear within other parentheses in numerical or algebraic expressions. Regardless of what symbol is used, ( ), { }, or [ ], perform operations in the innermost parentheses first and work outward.

The solutions are underlined.

401. The expression is the difference of two perfect squares.
The left side of the equation factors as $(x + 5)(x - 5) = 0$.
Applying the zero product property (if $(a)(b) = 0$, then $a = 0$ or $b = 0$ or both $= 0$), the first factor or the second factor or both must equal zero. $(x + 5) = 0$
Subtract 5 from both sides of the equation. $x + 5 - 5 = 0 - 5$
Combine like terms on each side. $x = {}^-5$
Let the second factor equal zero. $x - 5 = 0$
Add 5 to both sides of the equation. $x - 5 + 5 = 0 + 5$
Combine like terms on each side. $x = 5$
<u>The solutions for the equation are $x = 5$ and $x = {}^-5$.</u>

402. The expression is the difference of two perfect squares.
The left side of the equation factors as $(n + 13)(n - 13) = 0$.
Applying the zero product property (if $(a)(b) = 0$, then $a = 0$ or $b = 0$ or both $= 0$), the first factor or the second factor or both must equal zero. $(n + 13) = 0$
Subtract 13 from both sides of the equation. $n + 13 - 13 = 0 - 13$
Combine like terms on each side. $n = {}^-13$
Let the second factor equal zero. $n - 13 = 0$
Add 13 to both sides of the equation. $n - 13 + 13 = 0 + 13$
Combine like terms on each side. $n = 13$
<u>The solutions for the equation are $n = 13$ and $n = {}^-13$.</u>

403. Factor the trinomial expression on the left side using the trinomial factor form. $(a + 4)(a + 8) = 0$
Using the zero product property, subtract 4 from both sides. $(a + 4) = 0$
$a + 4 - 4 = 0 - 4$
Combine like terms on each side. $a = {}^-4$
Let the second factor equal zero. $(a + 8) = 0$

Subtract 8 from both sides. $a + 8 - 8 = 0 - 8$
Combine like terms on each side. $a = {}^-8$
The solutions for the equation are $a = {}^-4$ and $a = {}^-8$.

404. Factor the trinomial expression on the left side using the trinomial factor form. $(y - 8)(y - 7) = 0$
Using the zero product property, add 8 to both sides. $(y - 8) = 0$
$y - 8 + 8 = 0 + 8$
Combine like terms on each side. $y = 8$
Let the second factor equal zero. $(y - 7) = 0$
Add 7 to both sides. $y - 7 + 7 = 0 + 7$
Combine like terms on each side. $y = 7$
The solutions for the equation y are $y = 8$ and $y = 7$.

405. Factor the trinomial expression on the left side using the trinomial factor form. $(b + 10)(b - 9) = 0$
Using the zero product property, subtract 10 from both sides. $(b + 10) = 0$
$b + 10 - 10 = 0 - 10$
Combine like terms on each side. $b = {}^-10$
Let the second factor equal zero. $(b - 9) = 0$
Add 9 to both sides. $b - 9 + 9 = 0 + 9$
Combine like terms on each side. $b = 9$
The solutions for the equation are $b = {}^-10$ and $b = 9$.

406. Transform the equation so that all terms are on one side and zero is on the other side.
Subtract from both sides. $4x^2 - 49 = 49 - 49$
Combine like terms on each side. $4x^2 - 49 = 0$
The expression is the difference of two perfect squares.
The equation factors into $(2x + 7)(2x - 7) = 0.$
Applying the zero product property (if $(a)(b) = 0$, then $a = 0$ or $b = 0$ or both $= 0$), the first factor or the second factor or both must equal zero. $(2x + 7) = 0$
Subtract 7 from both sides of the equation. $2x + 7 - 7 = 0 - 7$
Combine like terms on each side. $2x = {}^-7$
Divide both sides by 2. $\frac{2x}{2} = \frac{^-7}{2}$
Simplify. $x = {}^-3\frac{1}{2}$
Let the second factor equal zero. $(2x - 7) = 0$
Add 7 to both sides of the equation. $2x - 7 + 7 = 0 + 7$
Combine like terms on both sides. $2x = 7$
Divide both sides by 2. $x = 3\frac{1}{2}$
The solutions for the equation are $x = {}^-3\frac{1}{2}$ and $x = 3\frac{1}{2}$.

407. Transform the equation so that all terms are on one side of the equation, and zero is on the other side. $81z^2 - 324 = 0$
A factor of 81 is common to both terms. Factor out 81 from both terms. $81(z^2 - 4) = 0$
Divide both sides by 81. $(z - 2)(z + 2) = 0$
The expression within the parentheses is the difference of two perfect squares. $81(z - 2)(z + 2) = 0$
Applying the zero product property (if $(a)(b) = 0$, then $a = 0$ or $b = 0$ or both $= 0$), the first factor or the second factor must equal zero. $(z - 2) = 0$
Add 2 to both sides of the equation. $z - 2 + 2 = 0 + 2$
Combine like terms on each side. $z = 2$
Let the second factor equal zero. $(z + 2) = 0$
Subtract 2 from both sides of the equation. $z + 2 - 2 = 0 - 2$
Combine like terms on each side. $z = {}^-2$
The solutions for the equation are $z = 2$ and $z = {}^-2$.

408. Factor the trinomial expression on the left side using the trinomial factor form. $(2n + 6)(n + 7) = 0$
Using the zero product property, subtract 6 from both sides. $(2n + 6) = 0$
$2n + 6 - 6 = 0 - 6$
Combine like terms on each side. $2n = {}^-6$
Divide both sides by 2. $\frac{2n}{2} = \frac{^-6}{2}$
Simplify terms. $n = {}^-3$
Let the second factor equal zero. $(n + 7) = 0$
Subtract 7 from both sides. $n + 7 - 7 = 0 - 7$
Combine like terms on each side. $n = {}^-7$
The solutions for the equation are $n = {}^-3$ and $n = {}^-7$.

409. Use the greatest common factor method to factor out 3 from all terms. $3(c^2 - 11c - 26) = 0$
Factor the trinomial expression on the left side using the trinomial factor form. $3(c - 13)(c + 2) = 0$
Ignore the factor 3 in the expression.
Using the zero product property, add 13 to both sides. $(c - 13) = 0$
$c - 13 + 13 = 0 +$
13
Combine like terms on each side. $c = 13$
Let the second factor equal zero. $(c + 2) = 0$
Subtract 2 from both sides. $c + 2 - 2 = 0 - 2$
Combine like terms on each side. $c = {}^-2$
The solutions for the equation are $c = 13$ and $c = {}^-2$.

410. Transform the equation so that all terms are on one side of the equation, and zero is on the other side. $6x^2 - 17x + 5 = 0$

Factor the trinomial expression on the left side using the trinomial factor form. $(3x - 1)(2x - 5) = 0$

Applying the zero product property (if $(a)(b) = 0$, then $a = 0$ or $b = 0$ or both = 0), the first factor or the second factor must equal zero. $(3x - 1) = 0$

Add 1 to both sides of the equation. $3x - 1 + 1 = 0 + 1$

Combine like terms on each side. $3 = 1$

Divide both sides of the equation by 3. $x = \frac{1}{3}$

Let the second factor equal zero. $(2x - 5) = 0$

Add 5 to both sides of the equation. $2x - 5 + 5 = 0 + 5$

Combine like terms on each side. $2x = 5$

Divide both sides of the equation by 2. $x = \frac{5}{2}$

The solutions for the equation are $x = \frac{1}{3}$ and $x = \frac{5}{2}$.

411. Use the greatest common factor method to factor out 3 from all terms. $3(x^2 - 12x + 36) = 0$

Factor the trinomial expression on the left side using the trinomial factor form. $3(x - 6)(x - 6) = 0$

Ignore the factor 3 in the expression.

Using the zero product property, add 6 to both sides. $(x - 6) = 0$

$x - 6 + 6 = 0 + 6$

Combine like terms on each side. $x = 6$

Since both factors of the trinomial expression are the same, the solution for the equation is $x = 6$.

412. Use the greatest common factor method to factor out 7 from all terms. $7(a^2 - 3a - 4) = 0$

Factor the trinomial expression on the left side using the trinomial factor form. $7(a - 4)(a + 1) = 0$

Ignore the factor 7 in the expression.

Using the zero product property, add 4 to both sides. $(a - 4) = 0$

$a - 4 + 4 = 0 + 4$

Combine like terms on each side. $a = 4$

Let the second factor equal zero. $(a + 1) = 0$

Subtract 1 from both sides. $a + 1 - 1 = 0 - 1$

Combine like terms on each side. $a = {}^{-}1$

The solutions for the equation are $a = {}^{-}1$ and $a = 4$.

413. Use the greatest common factor method. $8(y^2 + 7y + 12) = 0$
Factor the trinomial expression on the left side trinomial using the trinomial factor form. $8(y + 4)(y + 3) = 0$
Ignore the factor 8 in the expression.
Using the zero product property, subtract 4 from both sides. $(y + 4) = 0$
$y + 4 - 4 = 0 - 4$
Combine like terms on each side. $y = {}^-4$
Let the second factor equal zero. $(y + 3) = 0$
Subtract 3 from both sides. $y + 3 - 3 = 0 - 3$
Simplify. $y = {}^-3$
The solutions for the equation are $y = {}^-4$ and $y = {}^-3$.

414. Transform the equation into the familiar trinomial equation form.
Add 10 from both sides of the equation. $2x^2 + 9x + 10 = {}^-10 + 10$
Combine like terms on each side. $2x^2 + 9x + 10 = 0$
Factor the trinomial expression on the left side using the trinomial factor form. $(2x + 5)(x + 2) = 0$
Using the zero product property, subtract 5 from both sides. $(2x + 5) = 0$
$2x + 5 - 5 = 0 - 5$
Combine like terms on each side. $2x = {}^-5$
Divide both sides by 2. $\frac{2x}{2} = \frac{^-5}{2}$
Simplify terms. $x = {}^-2\frac{1}{2}$
Let the second term equal zero. $x + 2 = 0$
Subtract 2 to both sides. $x + 2 - 2 = 0 - 2$
Simplify. $x = {}^-2$
The solutions for the equation are $x = {}^-2\frac{1}{2}$ and $x = {}^-2$.

415. Transform the equation into the familiar trinomial equation form.

Subtract 15 from both sides of the equation. $4x^2 + 4x - 15 = 15 - 15$

Combine like terms on each side. $4x^2 + 4x - 15 = 0$

Factor the trinomial expression on the left side using the trinomial factor form. $(2x - 3)(2x + 5) = 0$

Using the zero product property, subtract 5 from both sides. $(2x + 5) = 0$

$2x + 5 - 5 = 0 - 5$

Simplify. $2x = {}^-5$

Divide both sides by 2. $\frac{2x}{2} = \frac{-5}{2}$

Simplify terms. $x = {}^-2\frac{1}{2}$

Let the second factor equal zero. $(2x - 3) = 0$

Add 3 to both sides. $2x - 3 + 3 = 0 + 3$

Simplify. $2x = 3$

Divide both sides by 2. $\frac{2x}{2} = \frac{3}{2}$

Simplify terms. $x = 1\frac{1}{2}$

The solutions for the quadratic equation $4x^2 + 4x = 15$ are $x = {}^-2\frac{1}{2}$ and $x = 1\frac{1}{2}$.

416. Transform the equation so that all terms are on one side of the equation, and zero is on the other side. $15x^2 - 35x + 20 = 0$

A factor of 5 is common to all three terms. Factor out 5 from all terms. $5(3x^2 - 7x + 4) = 0$

Divide both sides by 5. $3x^2 - 7x + 4 = 0$

Factor the trinomial expression on the left side using the trinomial factor form. $(3x - 4)(x - 1) = 0$

Applying the zero product property (if $(a)(b) = 0$, then $a = 0$ or $b = 0$ or both $= 0$), the first factor or the second factor must equal zero. $(3x - 4) = 0$

Add 4 to both sides of the equation. $3x - 4 + 4 = 0 + 4$

Combine like terms on each side. $3x = 4$

Divide both sides of the equation by 3. $x = \frac{4}{3}$

Let the second factor equal zero. $(x - 1) = 0$

Add 1 to both sides of the equation. $x - 1 + 1 = 0 + 1$

Combine like terms on each side. $x = 1$

The solutions for the equation are $x = \frac{4}{3}$ and $x = 1$.

417. Transform the equation into the familiar trinomial equation form.

Subtract $19x$ from both sides. $3x^2 - 19x = 19x - 19x - 20$

Combine like terms. $3x^2 - 19x = {}^-20$

Add 20 to both sides. $3x^2 - 19x + 20 = {}^-20 + 20$

Combine like terms on each side. $3x^2 - 19x + 20 = 0$

Factor the trinomial expression on the left side using the trinomial factor form. $(3x - 4)(x - 5) = 0$

Using the zero product property, add 4 to both sides. $(3x - 4) = 0$

$3x - 4 + 4 = 0 + 4$

Simplify. $3x = 4$

Divide both sides by 3. $\frac{3x}{3} = \frac{4}{3}$

Simplify terms. $x = 1\frac{1}{3}$

Now let the second term equal zero. $(x - 5) = 0$

Add 5 to both sides. $x - 5 + 5 = 0 + 5$

Simplify. $x = 5$

The solutions for the equation are $x = 1\frac{1}{3}$ and $x = 5$.

---

418. Transform the equation into the familiar trinomial equation form.

Subtract 42 from both sides of the equation. $8b^2 + 10b - 42 = 42 - 42$

Simplify. $8b^2 + 10b - 42 = 0$

Use the greatest common factor method to factor out 2. $2(4b^2 + 5b - 21) = 0$

Factor the trinomial expression on the left side using the trinomial factor form. $2(4b - 7)(b + 3) = 0$

Ignore the factor 2 in the expression.

Using the zero product property, add 7 to both sides. $(4b - 7) = 0$

$4b - 7 + 7 = 0 + 7$

Simplify. $4b = 7$

Divide both sides by 4. $\frac{4b}{4} = \frac{7}{4}$

Simplify terms. $b = 1\frac{3}{4}$

Now let the second term equal zero. $(b + 3) = 0$

Subtract 3 from both sides. $b + 3 - 3 = 0 - 3$

Simplify. $b = {}^-3$

The solutions for the equation are $b = 1\frac{3}{4}$ and $b = {}^-3$.

---

419. Transform the equation so that all terms are on one side of the equation, and zero is on the other side. $30x^2 - 17x + 1 = 0$

Factor the trinomial expression on the left side using the trinomial factor form. $(15x - 1)(2x - 1) = 0$

Applying the zero product property (if $(a)(b) = 0$, then $a = 0$ or $b = 0$ or both $= 0$), the first factor or the second factor must equal zero. $(15x - 1) = 0$

Add 1 to both sides of the equation. $15x - 1 + 1 = 0 + 1$

Combine like terms on each side. $15x = 1$

Divide both sides of the equation by 15. $x = \frac{1}{15}$

Let the second factor equal zero. $(2x - 1) = 0$

Add 1 to both sides of the equation. $2x - 1 + 1 = 0 + 1$

Combine like terms on each side. $2x = 1$

Divide both sides of the equation by 2. $x = \frac{1}{2}$

The solutions for the equation are $x = \frac{1}{15}$.

420. Transform the equation into the familiar trinomial equation form.

Add $9b$ to both sides of the equation. $6b^2 + 20b + 9b = {^{-}9b} + 9b - 20$

Simplify and add 20 to both sides of the equation. $6b^2 + 29b + 20 = 20 - 20$

Simplify. $6b^2 + 29b + 20 = 0$

Factor the trinomial expression using the trinomial factor form. $(6b + 5)(b + 4) = 0$

Using the zero product property, subtract 5 from both sides. $(6b + 5) = 0$

$6b + 5 - 5 = 0 - 5$

Simplify. $6b = {^{-}5}$

Divide both sides by 6. $\frac{6b}{6} = \frac{^{-}5}{6}$

Simplify terms. $b = \frac{^{-}5}{6}$

Now set the second factor equal to zero. $(b + 4) = 0$

Subtract 4 from both sides. $b + 4 - 4 = 0 - 4$

Simplify. $b = {^{-}4}$

The solutions for the equation are $b = \frac{^{-}5}{6}$ and $b = {^{-}4}$.

421. Factor the greatest common factor 5 from each term. $5(3x^2 - 14x - 24) = 0$

Now factor the trinomial expression on the left side using the trinomial factor form. $5(3x + 4)(x - 6) = 0$

Using the zero product property, subtract 4 from both sides. $(3x + 4) = 0$

$3x + 4 - 4 = 0 - 4$

Simplify. $3x = {}^-4$

Divide both sides by 3. $\frac{3x}{3} = \frac{-4}{3}$

Simplify terms. $x = \frac{-4}{3}$

Now set the second factor equal to zero. $(x - 6) = 0$

Add 6 to both sides. $x - 6 + 6 = 0 + 6$

Simplify. $x = 6$

The solutions for the equation are $x = \frac{-4}{3}$ and $x = 6$.

---

422. Transform the equation into the familiar trinomial equation form.

Subtract $52x$ from both sides of the equation. $7x^2 - 52x = 52x - 52x - 21$

Simplify and add 21 to both sides of the equation. $7x^2 - 52x + 21 = 21 - 21$

Simplify. $7x^2 - 52x + 21 = 0$

Factor the trinomial expression on the left side using the trinomial factor form. $(7x - 3)(x - 7) = 0$

Using the zero product property, add 3 to both sides. $(7x - 3) = 0$

$7x - 3 + 3 = 0 + 3$

Simplify. $7x = 3$

Divide both sides by 7. $\frac{7x}{7} = \frac{3}{7}$

Simplify terms. $x = \frac{3}{7}$

Now set the second factor equal to zero. $x - 7 = 0$

Add 7 to both sides of the equation. $x - 7 + 7 = 0 + 7$

Simplify. $x = 7$

The solutions for the equation are $x = \frac{3}{7}$ and $x = 7$.

---

423. Transform the equation into the familiar trinomial equation form.

Add 36 to both sides of the equation.	$36z^2 + 78z + 36 = {}^-36 + 36$
Combine like terms.	$36z^2 + 78z + 36 = 0$
Factor out the greatest common factor 6 from each term.	$6(6z^2 + 13z + 6) = 0$
Factor the trinomial expression on the left side into two factors.	$6(2z + 3)(3z + 2) = 0$
Ignore the numerical factor and set the first factor equal to zero.	$(2z + 3) = 0$
Subtract 3 from both sides.	$2z + 3 - 3 = 0 - 3$
Simplify terms.	$2z = {}^-3$
Divide both sides by 2.	$\frac{2z}{2} = \frac{^-3}{2}$
Simplify terms.	$z = {}^-1\frac{1}{2}$
Now let the second factor equal zero.	$(3z + 2) = 0$
Subtract 2 from both sides.	$3z + 2 - 2 = 0 - 2$
Simplify terms.	$3z = {}^-2$
Divide both sides by 3.	$\frac{3z}{3} = \frac{^-2}{3}$
Simplify terms.	$z = \frac{^-2}{3}$

The solutions for the equation are $z = {}^-1\frac{1}{2}$ and $z = \frac{^-2}{3}$.

424. the trinomial expression on the left side using the trinomial factor form.

the trinomial expression on the left side using the trinomial factor form.	$(20z - 3)(10z + 1) = 0$
Applying the zero product property (if $(a)(b) = 0$, then $a = 0$ or $b = 0$ or both = 0), the first factor or the second factor must equal zero.	$(20z - 3) = 0$
Add 3 to both sides of the equation.	$20z - 3 + 3 = 0 + 3$
Combine like terms on each side.	$20z = 3$
Divide both sides of the equation by 20.	$z = \frac{3}{20}$
Let the second factor equal zero.	$(10z + 1) = 0$
Subtract 1 from both sides of the equation.	$10z + 1 - 1 = 0 - 1$
Combine like terms on each side.	$10z = {}^-1$
Divide both sides of the equation by 10.	$z = {}^-\frac{1}{10}$

The solutions for the equation are $z = \frac{3}{20}$ and $z = {}^-\frac{1}{10}$.

425. Divide both sides of the equation by 3. $\frac{24x^2}{3} = \frac{3(43x - 15)}{3}$

Simplify terms. $8x^2 = 43x - 15$

Add $(15 - 43x)$ to both sides of the equation. $8x^2 + 15 - 43x = 43x - 15 + 15 - 43x$

Combine like terms. $8x^2 + 15 - 43x = 0$

Use the commutative property to move terms. $8x^2 - 43x + 15 = 0$

Factor the trinomial expression. $(8x - 3)(x - 5) = 0$

Using the zero product property, add 3 to both sides and divide by 8. $(8x - 3) = 0$

$\frac{8x}{8} = \frac{3}{8}$

Simplify terms. $x = \frac{3}{8}$

Now let the second factor equal zero. $(x - 5) = 0$

Add 5 to both sides. $x = 5$

The solutions for the equation are $x = \frac{3}{8}$ and $x = 5$.

18

Simplifying Radicals

This chapter will give you practice in operating with radicals. You will not always be able to factor polynomials by factoring whole numbers and whole number coefficients. Nor do all trinomials with whole numbers have whole numbers for solutions. In these last chapters, you will need to know how to operate with radicals.

The radical sign $\sqrt{\ }$ tells you to find the root of a number. The number under the radical sign is called the *radicand.*

Tips for Simplifying Radicals

Simplify radicals by completely factoring the radicand and taking out the square root. The most thorough method for factoring is to do a prime factorization of the radicand. Then you look for square roots that can be factored out of the radicand.

e.g., $\sqrt{28} = \sqrt{2 \cdot 2 \cdot 7} = \sqrt{2 \cdot 2} \cdot \sqrt{7} = 2\sqrt{7}$
or $\sqrt{4x^3} = \sqrt{2 \cdot 2 \cdot x \cdot x \cdot x} = \sqrt{2 \cdot 2} \cdot \sqrt{x \cdot x} \cdot \sqrt{x} = 2x\sqrt{x}$

You may also recognize perfect squares within the radicand. Then you can simplify their roots out of the radical sign.

If you get rid of the denominator within the radical sign, you will no longer have a fractional radicand. This is known as *rationalizing* the denominator.

e.g., $\sqrt{\frac{2x}{5}} = \sqrt{\frac{2x}{5} \cdot \frac{5}{5}} = \sqrt{\frac{10x}{5^2}} = \sqrt{\frac{1}{5^2} \cdot 10x} = \frac{1}{5}\sqrt{10x}$

When there is a radical in the denominator, you can rationalize the expression as follows:

$\frac{3}{\sqrt{6}} = \frac{3}{\sqrt{6}} \cdot \frac{\sqrt{6}}{\sqrt{6}} = \frac{3\sqrt{6}}{6} = \frac{1}{2}\sqrt{6}$

If the radicands are the same, radicals can be added and subtracted as if the radicals were variables.

e.g., $4\sqrt{3} + 2\sqrt{3} = 6\sqrt{3}$ or $5\sqrt{x} - 2\sqrt{x} = 3\sqrt{x}$

Product property of radicals

When multiplying radicals, multiply the terms in front of the radicals, then multiply the radicands and put that result under the radical sign.

e.g., $4\sqrt{3} \cdot 7\sqrt{5} = 4 \cdot 7\sqrt{3 \cdot 5} = 28\sqrt{15}$

Quotient property of radicals

When dividing radicals, first divide the terms in front of the radicals, and then divide the radicands and put that result under the radical sign.

e.g., $\frac{6\sqrt{10}}{3\sqrt{2}} = \frac{6}{3}\sqrt{\frac{10}{2}} = 2\sqrt{5}$

Simplify the following radical expressions. Be certain to rationalize all denominators.

426. $\sqrt{12}$

427. $\sqrt{500}$

428. $\sqrt{3n2}$

429. $\sqrt{24x5}$

430. $\sqrt{240x^3y^4z}$

431. $\sqrt{\frac{25w^6}{16}}$

432. $\sqrt{\frac{7}{4}}$

433. $\frac{3}{\sqrt{5}}$

434. $\frac{\sqrt{12xy}}{\sqrt{x}}$

435. $3\sqrt{3} + 6\sqrt{5} + 2\sqrt{5}$

436. $2\sqrt{7} - 3\sqrt{28}$

437. $(9\sqrt{a2b})(3a\sqrt{\text{b}})$

438. $2\sqrt{5} \cdot 3\sqrt{15}$

439. $\sqrt{\frac{81}{100} \cdot \frac{343}{z}}$

440. $\sqrt{\frac{8y^3z}{24yz^3}}$

441. $\frac{\sqrt{160}}{\sqrt{2}}$

442. $\frac{y}{3z} \cdot \sqrt{\frac{25z}{y^3}}$

443. $5\sqrt{\frac{8}{64}}$

444. $\frac{^-2\sqrt{128}}{\sqrt{2}}$

445. $\frac{\sqrt{27}}{\sqrt{72}}$

446. $\sqrt{\frac{8y}{w^3z}} \cdot \sqrt{\frac{w^3z}{64y}}$

447. $\sqrt{\frac{4}{3}} \cdot \sqrt{\frac{10}{3}}$

448. $\frac{\sqrt{15} \cdot \sqrt{105}}{^-3}$

449. $^-3 \cdot \sqrt{8} \cdot \sqrt{2} \cdot \sqrt{16}$

450. $\frac{\sqrt{6} \cdot \sqrt{4z} \cdot \sqrt{y}}{\sqrt{24y^3} \cdot \sqrt{y}}$

Answers

Numerical expressions in parentheses like this [ ] are operations performed on only part of the original expression. The operations performed within these symbols are intended to show how to evaluate the various terms that make up the entire expression.

Expressions with parentheses that look like this ( ) contain either numerical substitutions or expressions that are part of a numerical expression. Once a single number appears within these parentheses, the parentheses are no longer needed and need not be used the next time the entire expression is written.

When two pair of parentheses appear side by side like this ( )( ), it means that the expressions within are to be multiplied.

Sometimes parentheses appear within other parentheses in numerical or algebraic expressions. Regardless of what symbol is used, ( ), { }, or [ ], perform operations in the innermost parentheses first and work outward.

Underlined expressions show simplified result.

426. First, factor the radicand. $\sqrt{12} = \sqrt{2 \cdot 2 \cdot 3}$
Now take out the square root of any pair of factors or any perfect squares you recognize. $\underline{\sqrt{2 \cdot 2 \cdot 3} = 2\sqrt{3}}$

427. First, factor the radicand. $\sqrt{500} = \sqrt{5 \cdot 10 \cdot 10}$
Now take out the square root. $\underline{\sqrt{5 \cdot 10 \cdot 10} = 10\sqrt{5}}$

428. First, factor the radicand. $\sqrt{3n^2} = \sqrt{3 \cdot n \cdot n}$
Now take out the square root. $\underline{\sqrt{3 \cdot n \cdot n} = n\sqrt{3}}$

429. First, factor the radicand and look for squares. $\sqrt{24x^5} = \sqrt{6 \cdot 2 \cdot 2 \cdot x \cdot x \cdot x \cdot x \cdot x}$
Now take out the square root. $\underline{\sqrt{6 \cdot 2 \cdot 2 \cdot x \cdot x \cdot x \cdot x \cdot x} = 2x \cdot x\sqrt{6x} = 2x^2\sqrt{6x}}$

430. First, factor the radicand and look for squares. $\sqrt{2 \cdot 2 \cdot 2 \cdot 2 \cdot 3 \cdot 5 \cdot x \cdot x \cdot x \cdot y \cdot y \cdot y \cdot y \cdot z}$
Now, take out the square root. $\underline{2 \cdot 2 \cdot x \cdot y \cdot y \cdot \sqrt{3 \cdot 5 \cdot x \cdot z} = 4xy^2\sqrt{15xz}}$

431. First, factor the radicand and look for squares. $\sqrt{\frac{5 \cdot 5 \cdot w^3 \cdot w^3}{4 \cdot 4}}$
Use the laws of commutativity and associativity to rearrange the terms in the radicand. $\sqrt{\frac{5w^3}{4} \cdot \left(\frac{5w^3}{4}\right)}$
Now, take out the square root. $\frac{5w^3}{4}$

432. Use the quotient property of radicals and simplify the radicals in the numerator and denominator separately.

$$\sqrt{\frac{7}{4}} = \frac{\sqrt{7}}{\sqrt{4}} = \frac{\sqrt{7}}{\sqrt{2 \cdot 2}} = \frac{\sqrt{7}}{2}$$

433. For this expression, you must rationalize the denominator. Use the identity property of multiplication and multiply the expression by 1 in a form useful for your purposes. In this case, that is to get the radical out of the denominator.

$$\frac{3}{\sqrt{5}} = \frac{3}{\sqrt{5}} \cdot \frac{\sqrt{5}}{\sqrt{5}} = \frac{3\sqrt{5}}{5}$$

434. First, rationalize the denominator for this expression. Then see if it can be simplified any further. As in the previous problem, multiply the expression by 1 in a form suitable for this purpose.

$$\frac{\sqrt{12xy}}{\sqrt{x}} = \frac{\sqrt{12xy}}{\sqrt{x}} \cdot \frac{\sqrt{x}}{\sqrt{x}}$$

Use the product property of radicals to combine the radicands in the numerator. In the denominator, a square root times itself is the radicand by itself.

$$\frac{\sqrt{12xy}}{\sqrt{x}} \cdot \frac{\sqrt{x}}{\sqrt{x}} = \frac{\sqrt{12xyx}}{x}$$

Now factor the radicand.

$$\frac{\sqrt{12xyx}}{x} = \frac{\sqrt{3 \cdot 4x^2y}}{x}$$

Factoring out the square roots results in

$$\frac{\sqrt{3 \cdot 4x^2y}}{x} = \frac{2x\sqrt{3y}}{x}$$

The x in the numerator and the denominator divides out, leaving

$$2\sqrt{3y}.$$

435. In this expression, add the "like terms" as if the similar radicals were similar variables.

$$3\sqrt{3} + (6\sqrt{5} + 2\sqrt{5}) = 3\sqrt{3} + 8\sqrt{5}$$

The radical terms cannot be added further because the radicands are different.

436. Simplify the second term of the expression by factoring the radicand.

$$2\sqrt{7} - 3\sqrt{28} = 2\sqrt{7} - 3\sqrt{4 \cdot 7}$$

Now simplify the radicand.

$$2\sqrt{7} - 3\sqrt{4 \cdot 7} = 2\sqrt{7} - 3 \cdot 2\sqrt{7}$$

Finally, combine like terms.

$$2\sqrt{7} - 6\sqrt{7} = {}^{-}4\sqrt{7}$$

437. For this expression, use the product property of radicals and combine the factors in the radicand and outside the radical signs.

$$(9\sqrt{a^2b})(3a\sqrt{b}) = 9 \cdot 3a\sqrt{a^2b \cdot b} = 27a\sqrt{a^2b^2} = 27a \cdot ab = 27a^2b$$

438. Use the product property of radicals to simplify the expression.

$$2\sqrt{5} \cdot 3\sqrt{15} = 2 \cdot 3\sqrt{5 \cdot 15}$$

Now look for a perfect square in the radicand.

You can multiply and then factor or just factor first. $6\sqrt{5 \cdot 15} = 6\sqrt{75} = 6\sqrt{25 \cdot 3} = 6 \cdot 5\sqrt{3} = 30\sqrt{3}$

Or if you just factor the radicand, you will see the perfect square as 5 times 5. $6\sqrt{5 \cdot 15} = 6\sqrt{5 \cdot 5 \cdot 3} = 6 \cdot 5\sqrt{3} = \underline{30\sqrt{3}}$

439. First, factor the radicand and look for squares. $\sqrt{\frac{9 \cdot 9 \cdot 7 \cdot 7 \cdot 7}{10 \cdot 10 \cdot z}}$

Use the quotient property of radicals. $\frac{\sqrt{9 \cdot 9 \cdot 7 \cdot 7 \cdot 7}}{\sqrt{10 \cdot 10 \cdot z}}$

Now, take out the square root. $\frac{9 \cdot 7\sqrt{7}}{10\sqrt{z}} = \frac{63\sqrt{7}}{10\sqrt{z}}$

This result can be written a few different ways. $\frac{63\sqrt{7}}{10\sqrt{z}} = \frac{63}{10}\sqrt{\frac{7}{z}} = \frac{63}{10}\frac{\sqrt{7z}}{z} = \frac{63\sqrt{7z}}{10z}$

440. First, simplify the numerical coefficients within the radicand. $\sqrt{\frac{y^3z}{3yz^3}}$

When similar factors, or bases, are being divided, subtract the exponent in the denominator from the exponent in the numerator. $\sqrt{\frac{y^{3-1}z^{1-3}}{3}}$

Simplify the operations in the exponents. $\sqrt{\frac{y^2z^{-2}}{3}}$

A base with a negative exponent in the numerator is equivalent to the same variable or base in the denominator with the inverse sign for the exponent. $\sqrt{\frac{y^2}{3z^2}}$

Apply the quotient property for radicals. $\frac{\sqrt{y^2}}{\sqrt{3z^2}}$

Take out the square root. $\frac{y}{z\sqrt{3}}$

You can also rationalize the denominator to get an equivalent form of the simplified expression. $\underline{\frac{y}{z\sqrt{3}} = \frac{y}{z\sqrt{3}} \cdot \frac{\sqrt{3}}{\sqrt{3}} = \frac{y\sqrt{3}}{3z}}$

441. For this term, factor the radicand in the numerator and look for perfect squares.

$$\frac{\sqrt{160}}{\sqrt{2}} = \frac{\sqrt{4 \cdot 4 \cdot 2 \cdot 5}}{\sqrt{2}} = \frac{4\sqrt{2 \cdot 5}}{\sqrt{2}}$$

Use the product property in the numerator.

$$\frac{4\sqrt{2 \cdot 5}}{\sqrt{2}} = \frac{4\sqrt{2} \cdot \sqrt{5}}{\sqrt{2}}$$

Divide out the common factor of $\sqrt{2}$ in the numerator and the denominator.

$$\frac{4\sqrt{2} \cdot \sqrt{5}}{\sqrt{2}} = 4\sqrt{5}$$

442. Apply the quotient property for radicals.

$$\frac{y}{3z} \cdot \frac{\sqrt{25z}}{\sqrt{y^3}}$$

Factor each radicand.

$$\frac{y}{3z} \cdot \frac{\sqrt{5 \cdot 5 \cdot z}}{\sqrt{y \cdot y \cdot y}}$$

Take out the square root in the numerator and denominator.

$$\frac{y}{3z} \cdot \frac{5\sqrt{z}}{y\sqrt{y}}$$

Cancel a y that is common to both the numerator and denominator.

$$\frac{1}{3z} \cdot \frac{5\sqrt{z}}{\sqrt{y}} = \frac{5\sqrt{z}}{3z\sqrt{y}}$$

This result can be written a few different ways by using the quotient property of radicals, or by rationalizing the denominator.

$$\frac{5\sqrt{z}}{3z\sqrt{y}} = \frac{5}{3z}\sqrt{\frac{z}{y}} = \frac{5\sqrt{yz}}{3yz}$$

443. You use the quotient property to begin rationalizing the denominator.

$$5\sqrt{\frac{8}{64}} = 5\frac{\sqrt{8}}{\sqrt{64}}$$

The 5 becomes part of the numerator. Factor the numerator and simplify the perfect square in the denominator.

$$5\frac{\sqrt{8}}{\sqrt{64}} = \frac{5\sqrt{4 \cdot 2}}{8}$$

Simplify the numerator.

$$\frac{5\sqrt{4 \cdot 2}}{8} = \frac{5 \cdot 2\sqrt{2}}{8}$$

Divide out a common factor 2 from the numerator and denominator.

$$\frac{5 \cdot 2\sqrt{2}}{8} = \frac{5 \cdot 5\sqrt{2}}{4}$$

444. Begin simplifying this term by rationalizing the denominator.

$$\frac{^-2\sqrt{128}}{\sqrt{2}} \cdot \frac{\sqrt{2}}{\sqrt{2}} = \frac{^-2 \cdot \sqrt{128} \cdot \sqrt{2}}{\sqrt{2} \cdot \sqrt{2}}$$

Using the product property, simplify the numerator and write the product of the term in the denominator.

$$\frac{^-2 \cdot \sqrt{128} \cdot \sqrt{2}}{\sqrt{2} \cdot \sqrt{2}} = \frac{^-2 \cdot \sqrt{128 \cdot 2}}{2}$$

Divide out a common factor of 2 from the numerator and denominator.

$$^-\sqrt{128 \cdot 2}$$

Factor the radicand.

$$^-\sqrt{2 \cdot 2 \cdot 2 \cdot 2 \cdot 2 \cdot 2 \cdot 2 \cdot 2}$$

Now, take out the square root.

$$^-2 \cdot 2 \cdot 2 \cdot 2 = {}^-16$$

445. You will have to rationalize the denominator, but first factor the radicands and look for perfect squares.

$$\frac{\sqrt{27}}{\sqrt{72}} = \frac{\sqrt{9 \cdot 3}}{\sqrt{36 \cdot 2}} = \frac{3\sqrt{3}}{6\sqrt{2}}$$

You can simplify the whole numbers in the numerator and denominator by a factor of 3.

$$\frac{3\sqrt{3}}{6\sqrt{2}} = \frac{\sqrt{3}}{2\sqrt{2}}$$

Now rationalize the denominator.

$$\frac{\sqrt{3}}{2\sqrt{2}} \cdot \frac{\sqrt{2}}{\sqrt{2}} = \frac{\sqrt{3} \cdot \sqrt{2}}{2 \cdot \sqrt{2} \cdot \sqrt{2}}$$

Simplify terms by using the product property in the numerator and multiplying terms in the denominator.

$$\frac{\sqrt{3} \cdot \sqrt{2}}{2 \cdot \sqrt{2} \cdot \sqrt{2}} = \frac{\sqrt{3 \cdot 2}}{2 \cdot 2} = \frac{\sqrt{6}}{4}$$

446. Apply the quotient property of radicals.

$$\frac{\sqrt{8y}}{\sqrt{w^3z}} \cdot \frac{\sqrt{w^3z}}{\sqrt{64y}}$$

Cancel the factor $\sqrt{w^3z}$ that is common to both the numerator and denominator.

$$\frac{\sqrt{8y}}{\sqrt{64y}}$$

Apply the quotient property of radicals.

$$\sqrt{\frac{8y}{64y}}$$

Cancel like factors in the numerator and denominator of the radicand.

$$\sqrt{\frac{8y}{64y}} = \sqrt{\frac{8y}{8(8y)}} = \sqrt{\frac{1}{8}}$$

Factor the radicand.

$$\sqrt{\frac{1}{2} \cdot \frac{1}{2} \cdot \frac{1}{2}}$$

Take out the square root.

$$\frac{1}{2}\sqrt{\frac{1}{2}}$$

This result can be written a few different ways.

$$\underline{\frac{1}{2}\sqrt{\frac{1}{2}} = \frac{1}{2\sqrt{2}} = \frac{\sqrt{2}}{4}}$$

447. Using the product property, put all terms in one radical sign.

$$\sqrt{\frac{4}{3}} \cdot \sqrt{\frac{10}{3}} = \sqrt{\frac{4 \cdot 10}{3 \cdot 3}}$$

Now use the quotient property to continue simplifying.

$$\sqrt{\frac{4 \cdot 10}{3 \cdot 3}} = \frac{\sqrt{4 \cdot 10}}{\sqrt{9}}$$

Simplify the perfect squares in the numerator and denominator.

$$\frac{\sqrt{4 \cdot 10}}{\sqrt{9}} = \frac{2\sqrt{10}}{3}$$

The expression is fine the way it is, or it could be written as

$$\underline{\tfrac{2}{3}\sqrt{10}.}$$

448. Begin by factoring the radicand of the second radical.
$\frac{\sqrt{15} \cdot \sqrt{105}}{^-3} = \frac{\sqrt{15} \cdot \sqrt{15 \cdot 7}}{^-3}$

Now use the product property to separate the factors of the second radical term into two radical terms. Why? Because you will then have the product of two identical radicals.
$\frac{\sqrt{15} \cdot \sqrt{15 \cdot 7}}{^-3} = \frac{\sqrt{15} \cdot \sqrt{15 \cdot 7}}{^-3} = \frac{\sqrt{15 \cdot 7}}{^-3}$

Now simplify the whole numbers.
$\frac{15\sqrt{7}}{^-3} = {^-5}\sqrt{7}$

449. Apply the product property of radicals to combine the three radical terms.
$^-3 \cdot \sqrt{2 \cdot 8 \cdot 16}$

Simplify.
$^-3 \cdot \sqrt{16 \cdot 16}$

Take out the square root.
$^-3(16)$

Simplify.
$\underline{^-48}$

450. Apply the product rule for radicals in both the numerator and denominator.
$\frac{\sqrt{24z \cdot y}}{\sqrt{24y^3 \cdot z^5}}$

Apply the quotient rule for radicals.
$\sqrt{\frac{24z \cdot y}{24y^3 \cdot z^5}}$

First, simplify the numerical coefficients within the radicand.
$\sqrt{\frac{z \cdot y}{y^3 \cdot z^5}}$

When similar factors, or bases, are being divided, subtract the exponent in the denominator from the exponent in the numerator.
$\sqrt{z^{1-5} \cdot y^{1-3}}$

Simplify the operations in the exponents.
$\sqrt{z^{^-4} \cdot y^{^-2}}$

A base with a negative exponent in the numerator is equivalent to the same variable or base in the denominator with the inverse sign for the exponent.
$\sqrt{\frac{1}{z^4y^2}}$

Apply the quotient property for radicals and use the fact that $\sqrt{1} = 1$.
$\sqrt{\frac{1}{z^4y^2}}$

Factor the radicand in the denominator.
$\frac{1}{\sqrt{z^2 \cdot z^2 \cdot y \cdot y}}$

Take out the square root.
$\frac{1}{z^2y}$

19

Solving Radical Equations

This chapter will give you more practice operating with radicals. However, the focus here is to use radicals to solve equations. An equation is considered a *radical equation* when the radicand contains a variable. When you use a radical to solve an equation, you must be aware of the positive and negative roots. You should always check your results in the original equation to see that both solutions work. When one of the solutions does not work, it is called an *extraneous solution*. When neither solution works in the original equation, there is said to be no solution.

Tips for Solving Radical Equations

- Squaring both sides of an equation is a valuable tool when solving radical equations. Use the following property: When a and b are algebraic expressions, if $a = b$, then $a^2 = b^2$.
- Isolate the radical on one side of an equation before using the squaring property.
- Squaring a radical results in the radical symbol disappearing, e.g., $(\sqrt{x+5})^2 = x + 5$.
- For second-degree equations, use the radical sign on both sides of the equation to find a solution for the variable. Check your answers.

Solve the following radical equations. Watch for extraneous solutions.

451. $x^2 = 49$

452. $x^2 = 135$

453. $\sqrt{n} = 11$

454. $2\sqrt{a} = 24$

455. $\sqrt{2x} - 4 = 4$

456. $\sqrt{4x + 6} = 8$

457. $\sqrt{3x + 4} + 8 = 12$

458. $2\sqrt{4x + 5} + 4 = 14$

459. $\sqrt{4x + 9} = {}^{-}13$

460. $\sqrt{9 - x} + 14 = 25$

461. $\surd 3\sqrt{6 - 2x} - 9 = 21$

462. $3\sqrt{3x + 1} = 15$

463. $3\sqrt{{}^{-}x} + 7 = 25$

464. $3 = 10 - \sqrt{100x - 1}$

465. $5\sqrt{{}^{-}1 - 3x} - 32 = {}^{-}17$

466. ${}^{-}7 = 10 - \sqrt{25x + 39}$

467. $3\sqrt{13x + 43} - 4 = 29$

468. $x = \sqrt{8 - 2x}$

469. $x = \sqrt{3x + 4}$

470. $x = \sqrt{7x - 10}$

471. $4x = \sqrt{2x^2 + 56}$

472. $3x = \sqrt{10 - x^2}$

473. $\sqrt{4x + 3} = 2x$

474. $\sqrt{2 - \frac{7}{2}x} = x$

475. $x = \sqrt{\frac{3}{2}x + 10}$

Answers

Numerical expressions in parentheses like this [ ] are operations performed on only part of the original expression. The operations performed within these symbols are intended to show how to evaluate the various terms that make up the entire expression.

Expressions with parentheses that look like this ( ) contain either numerical substitutions or expressions that are part of a numerical expression. Once a single number appears within these parentheses, the parentheses are no longer needed and need not be used the next time the entire expression is written.

When two pair of parentheses appear side by side like this ( )( ), it means that the expressions within are to be multiplied.

Sometimes parentheses appear within other parentheses in numerical or algebraic expressions. Regardless of what symbol is used, ( ), { }, or [ ], perform operations in the innermost parentheses first and work outward.

The solution is underlined.

451. Use the radical sign on both sides of the equation. $\sqrt{x^2} = \sqrt{49}$

Show both solutions for the square root of 49. $x = \pm 7$

Check the first solution in the original equation. $(7)^2 = 49$

$49 = 49$

Check the second solution in the original equation. $(^-7)^2 = 49$

$49 = 49$

Both solutions, $\underline{x = \pm 7}$, check out.

452. Use the radical sign on both sides of the equation. $\sqrt{x^2} = \sqrt{135}$

Simplify the radical. $x = \pm\sqrt{135} = \pm\sqrt{9 \cdot 15} = \pm 3\sqrt{15}$

Check the first solution in the original equation. $(3\sqrt{15})^2 = 135$

$3^2(\sqrt{15})^2 = 135$

$9(15) = 135$

$135 = 135$

Check the second solution in the original equation. $(^-3\sqrt{15})^2 = 135$

$(^-3)^2(\sqrt{15})^2 = 135$

$(9)(15) = 135$

$135 = 135$

Both solutions, $\underline{x = \pm 3\sqrt{15}}$, check out.

453. First, square both sides of the equation. $(\sqrt{n})^2 = 11^2$

Simplify both terms. $n = 121$

Check by substituting in the original equation. $\sqrt{121} = 11$

The original equation asks for only the positive root of n. So when you substitute 121 into the original equation, only the positive root $\underline{\sqrt{n} = 11}$ is to be considered. 11 = 11 checks out. Although this may seem trivial at this point, as the radical equations become more complex, this will become important.

454. Isolate the radical on one side of the equation. Divide both sides by 2. $\frac{2\sqrt{a}}{2} = \frac{24}{2}$

Simplify terms. $\sqrt{a} = 12$

Now square both sides of the equation. $(\sqrt{a})^2 = 12^2$

Simplify terms. $a = 144$

Check the solution in the original equation. $2\sqrt{144} = 24$

$2(12) = 24$

The solution $\underline{a = 144}$ checks out. $24 = 24$

455. Begin by adding 4 to both sides to isolate the radical. $\sqrt{2x} - 4 + 4 = 4 + 4$

Combine like terms on each side. $\sqrt{2x} = 8$

Square both sides of the equation. $(\sqrt{2x})^2 = 8^2$

Simplify terms. $2x = 64$

Divide both sides by 2. $x = 32$

Check the solution in the original equation. $\sqrt{2(32)} - 4 = 4$

$\sqrt{64} - 4 = 4$

The solution $\underline{x = 32}$ checks out. $8 - 4 = 4, 4 = 4$

456. Square both sides of the equation. $(\sqrt{4x + 6})^2 = 8^2$

$4x + 6 = 64$

Subtract 6 from both sides of the equation. $4x = 58$

Divide both sides by 4 and simplify. $x = \frac{58}{4} = 14.5$

Check the solution in the original equation. $\sqrt{4(14.5) + 6} = 8$

Simplify terms. $\sqrt{64} = 8$

The solution $\underline{x = 14.5}$ checks out. $8 = 8$

457. Subtract 8 from both sides in order to isolate the radical. $\sqrt{3x + 4} + 8 - 8 = 12 - 8$

$\sqrt{3x + 4} = 4$

Square both sides of the equation. $(\sqrt{3x + 4})^2 = 4^2$

Simplify terms. $3x + 4 = 16$

$3x + 4 - 4 = 16 - 4$

$3x = 12$

Subtract 4 from both sides and divide by 3. $x = 4$

Check your solution in the original equation. $\sqrt{3(4) + 4} + 8 = 12$

Simplify terms. $\sqrt{16} + 8 = 12$

$4 + 8 = 12$

The solution $\underline{x = 4}$ checks out. $12 = 12$

458. Subtract 4 from both sides of the equation. $2\sqrt{4x+5} = 10$

Divide both sides of the equation by 2 to completely isolate the radical on one side of the equation. $\sqrt{4x+5} = 5$

Square both sides of the equation. $(\sqrt{4x+5})^2 = (5)^2$

Simplify terms on both sides of the equation. $4x + 5 = 25$

Subtract 5 from both sides of the equation, and then divide by 4. $x = 5$

Check your solution in the original equation. $2\sqrt{4(5)+5} + 4 = 14$

Simplify terms under the radical sign. $2\sqrt{25} + 4 = 14$

Find the positive square root of 25. $2(5) + 4 = 14$

Simplify. $14 = 14$

The solution $\underline{x = 5}$ checks out.

459. Square both sides of the equation. $(\sqrt{4x+9})^2 = (^-13)^2$

Simplify terms on both sides of the equation. $4x + 9 = 169$

Subtract 9 from both sides and then divide by 4. $x = 40$

Substitute the solution in the original equation. $\sqrt{4(40+9)} = {}^-13$

Simplify the expression under the radical sign. $\sqrt{169} = {}^-13$

The radical sign calls for the positive square root. $13 \neq {}^-13$

The solution does not check out.

<u>There is no solution for this equation.</u>

460. Subtract 14 from both sides to isolate the radical. $\sqrt{9-x} = 11$

Now square both sides of the equation. $9 - x = 121$

Subtract 9 from both sides. $^-x = 112$

Multiply both sides by negative 1 to solve for x. $x = {}^-112$

Check the solution in the original equation. $\sqrt{9-(^-112)} + 14 = 25$

Simplify the expression under the radical sign. $\sqrt{121} + 14 = 25$

The square root of 121 is 11. Add 14 and the solution $\underline{x = {}^-112}$ checks out. $25 = 25$

461. Add 9 to both sides of the equation. $3\sqrt{6-2x} = 30$

Divide both sides of the equation by 3 to completely isolate the radical on one side of the equation. $\sqrt{6-2x} = 10$

Square both sides of the equation. $(\sqrt{6-2x})^2 = (10)^2$

Simplify terms on both sides of the equation. $6 - 2x = 100$

Subtract 6 from both sides of the equation, and then divide by -2. $x = {}^-47$

Check your solution in the original equation. $3\sqrt{6-2(-47)} - 9 = 21$

Simplify terms under the radical sign. $3\sqrt{100} - 9 = 21$

Find the positive square root of 100. $3(10) - 9 = 21$

Simplify. $21 = 21$

The solution $\underline{x = {}^-47}$ checks out.

462. To isolate the radical, divide both sides by 3. $\sqrt{3x+1} = 5$
Square both sides of the equation. $(\sqrt{3x+1})^2 = 5^2$
Simplify terms. $3x + 1 = 25$
Subtract 1 from both sides of the equation and divide by 3. $x = 8$
Check the solution in the original equation. $3\sqrt{13(8)+1} = 15$
Simplify the expression under the radical sign. $3\sqrt{25} = 15$
Multiply 3 by the positive root of 25. $3(5) = 15$
The solution $\underline{x = 8}$ checks out. $15 = 15$

463. Subtract 7 from both sides of the equation. $3\sqrt{^-x} = 18$
Divide both sides by 3 to isolate the radical. $\sqrt{^-x} = 6$
Square both sides of the equation. $(\sqrt{^-x})^2 = 6^2$
Simplify terms. $^-x = 36$
Multiply both sides by negative 1. $x = {^-36}$
Check the solution in the original equation. $3\sqrt{^-(^-36)} + 7 = 25$
Simplify terms under the radical. $3\sqrt{36} + 7 = 25$
Use the positive square root of 25. $3(6) + 7 = 25$
$18 + 7 = 25$
The solution $\underline{x = {^-36}}$ checks out. $25 = 25$

464. Add $\sqrt{100x-1}$ to both sides of the equation. $3 + \sqrt{100x-1} = 10 - \sqrt{100x-1} + \sqrt{100x-1}$
Combine like terms and simplify the equation. $3 + \sqrt{100x-1} = 10$
Now subtract 3 from both sides. $\sqrt{100x-1} = 7$
Square both sides of the equation. $100x - 1 = 49$
Add 1 to both sides of the equation and divide by 100. $x = 0.5$
Check the solution in the original equation. $3 = 10 - \sqrt{100(0.5) - 1}$
Simplify the expression under the radical sign. $3 = 10 - \sqrt{49}$
The equation asks you to subtract the positive square root of 49 from 10. $3 = 10 - 7$
The solution $\underline{x = 0.5}$ checks out. $3 = 3$

465. Add 32 to both sides of the equation. $5\sqrt{^-1 - 3x} = 15$
Divide both sides of the equation by 5 to completely isolate the radical on one side of the equation. $\sqrt{^-1 - 3x} = 3$
Square both sides of the equation. $(\sqrt{^-1 - 3x})^2 = (3)^2$
Simplify terms on both sides of the equation. $^-1 - 3x = 9$
Add 1 to both sides of the equation, and then divide by $^-3$. $x = \frac{-10}{3}$
Check your solution in the original equation. $5\sqrt{^-1 - 3(\frac{-10}{3})} - 32 = {^-17}$
Simplify terms under the radical sign. $5\sqrt{9} - 32 = {^-17}$

Find the positive square root of 9. $5(3) - 32 = {}^-17$
Simplify. ${}^-17 = {}^-17$
The solution $x = \frac{-10}{3}$ checks out.

466. Add $\sqrt{25x + 39}$ to both sides of the equation. $\sqrt{25x + 39} - 7 = 10 - \sqrt{25x + 39} + \sqrt{25x + 39}$
Combine like terms and simplify the equation. $\sqrt{25x + 39} - 7 = 10$
Add 7 to both sides of the equation. $\sqrt{25x + 39} = 17$
Square both sides. $25x + 39 = 289$
Subtract 39 from both sides and divide the result by 25. $x = 10$
Check the solution in the original equation. ${}^-7 = 10 - \sqrt{25(10) + 39}$
Simplify the expression under the radical sign. ${}^-7 = 10 - \sqrt{289}$
The equation asks you to subtract the positive square root of 289 from 10. ${}^-7 = 10 - 17$
The solution $\underline{x = 10}$ checks out. ${}^-7 = {}^-7$

467. To isolate the radical on one side of the equation, add 4 to both sides and divide the result by 3. $\sqrt{13x + 43} = 11$
Square both sides of the equation. $13x + 43 = 121$
Subtract 43 from both sides and divide by 13. $x = 6$
Check the solution in the original equation. $3\sqrt{13(6) + 43} - 4 = 29$
Simplify the expression under the radical sign. $3\sqrt{121} - 4 = 29$
Evaluate the left side of the equation. $3(11) - 4 = 29$
The solution $\underline{x = 6}$ checks out. $29 = 29$

468. The radical is alone on one side. Square both sides. $x^2 = 8 - 2x$
Transform the equation by putting all terms on the left side. $x^2 + 2x - 8 = 0$
The result is a quadratic equation. Solve for x by factoring using the trinomial factor form and setting each factor equal to zero and solving for x. (Refer to Chapter 16 for practice and tips for factoring quadratic equations.) It will be important to check each solution. $x^2 + 2x - 8 = (x + 4)(x - 2) = 0$
Let the first factor equal zero and solve for x. $x + 4 = 0$
Subtract 4 from both sides. $x = {}^-4$
Check the solution in the original equation. $({}^-4) = \sqrt{8 - 2({}^-4)}$
Evaluate the expression under the radical sign. ${}^-4 = \sqrt{16}$
The radical sign calls for a positive root. ${}^-4 \neq 4$

Therefore, x cannot equal $^-4$. $x = {}^-4$ is an example of an *extraneous root.*

Let the second factor equal zero and solve for x. $x - 2 = 0$

Subtract 2 from both sides. $x = 2$

Check the solution in the original equation. $(2) = \sqrt{8 - 2(2)}$

Evaluate the expression under the radical sign. $2 = \sqrt{4}$

The positive square root of 4 is 2. $2 = 2$

Therefore, the only solution for the equation is $\underline{x = 2.}$

469. With the radical alone on one side of the equation, square both sides. $x^2 = 3x + 4$

The resulting quadratic equation may have up to two solutions. Put it into standard form and factor the equation using the trinomial factor form to find the solutions. Then check the solutions in the original equation. $x^2 - 3x - 4 = (x - 4)(x + 1) = 0$

Letting each factor equal zero and solving for x results in two possible solutions, $x = 4$ and/or $^-1$. Check the first possible solution in the original equation. $(4) = \sqrt{3(4) + 4} = \sqrt{16} = 4$

The solution checks out. Now check the second possible solution in the original equation. $({}^-1) = \sqrt{3({}^-1) + 4} = \sqrt{1} = 1$

$^-1 \neq 1$

Therefore, $x = {}^-1$ is an *extraneous root.*

The only solution for the original equation is $\underline{x = 4.}$

470. Square both sides of the equation. $x^2 = 7x - 10$

Subtract $(7x - 10)$ from both sides of the equation. $x^2 - 7x + 10 = 0$

Factor the quadratic equation to find the solutions, and check each in the original equation to rule out any extraneous solution. $x^2 - 7x + 10 = (x - 5)(x - 2) = 0$

The first factor will give you the solution $x = 5$. The second factor will give the solution $x = 2$. Check the first solution for the quadratic equation in the original equation. $(5) = \sqrt{7(5) - 10}$

Simplify the expression under the radical sign. $5 = \sqrt{25}$ or $5 = 5$

The solution $\underline{x = 5}$ is a solution to the original equation.

Now check the second solution to the quadratic equation in the original. $(2) = \sqrt{7(2) - 10}$

Simplify the expression under the radical sign. $2 = \sqrt{4}$ or $2 = 2$

There are two solutions to the original equation, $\underline{x = 2 \text{ and } x = 5.}$

471. Square both sides of the equation. $(4x)^2 = (\sqrt{2x^2 + 56})^2$

Simplify terms on both sides of the equation. $16x^2 = 2x^2 + 56$

Subtract $2x^2$ from both sides of the equation, and then divide by 14. $x^2 = 4$

Transform the equation by putting all terms on the left side. $x^2 - 4 = 0$

The expression on the left side is a difference of two perfect squares. $(x - 2)(x + 2) = 0$

Set each factor equal to zero to get the possible solutions. $x = -2, x = 2$

Check both of these solutions in the original equation.

$\underline{x = -2:}$

$4(-2) = \sqrt{2(-2)^2 + 56}$

$-8 = \sqrt{64}$

$-8 = 8$ FALSE!

$\underline{x = 2:}$

$4(2) = \sqrt{2(2)^2 + 56}$

$8 = \sqrt{64}$

$8 = 8$ TRUE!

We conclude that $x = {}^-8$ is not a solution, but the solution $\underline{x = 8}$ checks out.

472. Square both sides of the equation. $(3x)^2 = (\sqrt{10 - x^2})^2$

Simplify terms on both sides of the equation. $9x^2 = 10 - x^2$

Add x^2 to both sides of the equation, and then divide by 10. $x^2 = 1$

Transform the equation by putting all terms on the left side. $x^2 - 1 = 0$

The expression on the left side is a difference of two perfect squares. $(x - 1)(x + 1) = 0$

Set each factor equal to zero to get the possible solutions. $x = {}^-1, x = 1$

Check both of these solutions in the original equation.

$\underline{x = {}^-1:}$
$3({}^-1) = \sqrt{10 - ({}^-1)^2}$
${}^-3 = \sqrt{9}$
${}^-3 = 3$ FALSE!
$\underline{x = 1:}$
$3(1) = \sqrt{10 - (1)^2}$
$3 = \sqrt{9}$
$3 = 3$ TRUE!

We conclude that $x = {}^-1$ is not a solution, but the solution $x = 1$ checks out.

473. Square both sides of the radical equation. $4x + 3 = 4x^2$

Transform the equation into a quadratic equation. $4x^2 - 4x - 3 = 0$

Factor the result using the trinomial factor form. $4x^2 - 4x - 3 = (2x + 1)(2x - 3) = 0$

Let the first factor equal zero and solve for x. $2x + 1 = 0$, so $x = {}^-0.5$

Let the second factor equal zero and solve for x. $2x - 3 = 0$, so $x = 1.5$

When you substitute ${}^-0.5$ for x in the original equation, the result will be $\sqrt{1} = {}^-1$. That cannot be true for the original equation, so $x = {}^-0.5$ is an extraneous root.

Substitute 1.5 for x in the original equation. $\sqrt{4(1.5) + 3} = 2(1.5)$

Simplify the terms on each side of the equal sign. $\sqrt{9} = 3$ or $3 = 3$

The only solution for the original equation is $\underline{x = 1.5}$.

474. Square both sides of the equation. $2 - \frac{7}{2}x = x^2$

Subtract $(2 - \frac{7}{2}x)$ from both sides of the equation. $0 = x^2 + \frac{7}{2}x - 2$

Multiply both sides of the equation by 2 to simplify the fraction. $0 = 2x^2 + 7x - 4$

Factor using the trinomial factor form. $2x^2 + 7x - 4 = (2x - 1)(x + 4) = 0$

Let the first factor equal zero and solve for x. $2x - 1 = 0$, so $x = \frac{1}{2}$

Let the second factor equal zero and solve for x. $x + 4 = 0$, so $x = {}^-4$

Check the first possible solution in the original equation. $\sqrt{2 - \frac{7}{2}(\frac{1}{2})} = (\frac{1}{2})$

Simplify the expression under the radical sign. $\sqrt{2 - \frac{7}{4}} = \sqrt{2 - 1\frac{3}{4}} = \sqrt{\frac{1}{4}} = \frac{1}{2}$

$\frac{1}{2} = \frac{1}{2}$

So $x = \frac{1}{2}$ is a solution.

Check the solution $x = {}^-4$ in the original equation. $\sqrt{2 - \frac{7}{2}({}^-4)} = ({}^-4)$

Simplify the expression. $\sqrt{16} = ({}^-4)$ or $4 = {}^-4$

This is not true, so $x = {}^-4$ is an extraneous root. There is one solution for the original equation, $x = \frac{1}{2}$.

475. Square both sides of the equation. $x^2 = \frac{3}{2}x + 10$

Add $({}^-\frac{3}{2}x - 10)$ to both sides of the equation. $x^2 - \frac{3}{2}x - 10 = 0$

Multiply the equation by 2 to eliminate the fraction. $2x^2 - 3x - 20 = 0$

Factor using the trinomial factor form. $2x^2 - 3x - 20 = (x - 4)(2x + 5) = 0$

Letting each factor of the trinomial factors equal zero results in two possible solutions for the original equation, $x = 4$ and/or $x = {}^-2\frac{1}{2}$.

Check the first possible solution in the original equation. $4 = \sqrt{\frac{3}{2}(4) + 10}$

Simplify the radical expression. $4 = \sqrt{16}$ or $4 = 4$

The solution $x = 4$ checks out as a solution for the original equation.

Check the second possible solution in the original equation. ${}^-2\frac{1}{2} = \sqrt{\frac{3}{2}({}^-2\frac{1}{2}) + 10}$

A negative number cannot be equal to a positive square root as the radical sign in the original expression calls for. Therefore, $x = {}^-2\frac{1}{2}$ is **not** a solution to the original equation. The only solution for this equation is $x = 4$.

20

Solving Equations with the Quadratic Formula

In this chapter, you will have the opportunity to practice solving equations using the quadratic formula. In Chapter 17, you practiced using factoring to solve quadratic equations, but factoring is useful only for those equations that can easily be factored. The quadratic formula will allow you to find solutions for any quadratic equation that can be put in the form $ax^2 + bx + c = 0$, where a, b, and c are numbers and $a \neq 0$.

The quadratic formula tells you that for any equation in the form $ax^2 + bx + c = 0$, the solution will be $x = \frac{^-b \pm \sqrt{b^2 - 4ac}}{2a}$. The solutions found using the quadratic formula are also called the roots of the equation. Some solutions will be in the form of whole numbers or fractions. Some will be in the form of a radical. Some will be undefined, as when the radicand is equal to a negative number. The ± in the quadratic equation tells you that there will be two solutions, one when you add the radical and one when you subtract it.

Tips for Solving Equations with the Quadratic Formula

- Transform the equation into the form $ax^2 + bx + c = 0$. Use the values for a, b, and c in the quadratic equation to determine the solution for the original equation.
- For solutions that contain a radical, be sure to simplify the radical as you practiced in Chapter 18.
- When you are asked to find the solution to the nearest hundredth, you can use a calculator to estimate the value of the radical.

Solve the following equations using the quadratic formula. Reduce answers to their simplest form or to the simplest radical form.

476. $x^2 + 2x - 8 = 0$

477. $2x^2 - 7x - 30 = 0$

478. $6x^2 + 13x - 28 = 0$

479. $18x^2 + 9x + 1 = 0$

480. $36x^2 - 8x = 6 - 27x$

481. $14x^2 = 12x + 32$

482. $4x^2 + 5x = 0$

483. $5x^2 = 27$

484. $5x^2 = 18x - 17$

485. $9x^2 - 30x + 25 = 0$

486. $^{-}12x^2 + 2 = {}^{-}5x$

487. $x^2 = \frac{^{-}5x - 2}{2}$

488. $x^2 + 8x = 5$

489. $1 - x = 56x^2$

490. $23x^2 = 2(8x - 1)$

491. $x^2 + 10x + 11 = 0$

492. $24x^2 + 18x - 6 = 0$

493. $7x^2 = 4(3x + 1)$

494. $\frac{1}{3}x^2 + \frac{3}{4}x - 3 = 0$

495. $0.5x^2 + 1.5x - 2 = 0$

Find the solutions to the following equations to the nearest hundredth.

496. $11r^2 - 4r - 7 = 0$

497. $3m^2 + 21m - 8 = 0$

498. $4y^2 = 16y - 5$

499. $5s^2 + 12s - 1 = 0$

500. $4z^2 - 11z + 2 = 0$

501. $11k^2 - 32k + 10 = 0$

Answers

Numerical expressions in parentheses like this [ ] are operations performed on only part of the original expression. The operations performed within these symbols are intended to show how to evaluate the various terms that make up the entire expression.

Expressions with parentheses that look like this ( ) contain either numerical substitutions or expressions that are part of a numerical expression. Once a single number appears within these parentheses, the parentheses are no longer needed and need not be used the next time the entire expression is written.

When two pair of parentheses appear side by side like this ( )( ), it means that the expressions within are to be multiplied.

Sometimes parentheses appear within other parentheses in numerical or algebraic expressions. Regardless of what symbol is used, ( ), { }, or [ ], perform operations in the innermost parentheses first and work outward.

The solutions are underlined.

476. The equation is already in the proper form.

First, list the values for a, b, and c. $a = 1 \quad b = 2 \quad c = {}^-8$

Substitute the values into the quadratic formula. $x = \frac{^-2 \pm \sqrt{(2)^2 - 4(1)(^-8)}}{2}$

Simplify the expression under the radical sign. $x = \frac{^-2 \pm \sqrt{4 - {}^-32}}{2} = \frac{^-2 \pm \sqrt{36}}{2}$

Evaluate the square root of 36. $x = \frac{^-2 \pm 6}{2}$

Find the two solutions for x by simplifying terms. First add the terms in the numerator, and then subtract them.

$x = \frac{^-2 + 6}{2} = \frac{4}{2} = 2$ and

$x = \frac{^-2 - 6}{2} = \frac{^-8}{2} = {}^-4$

The two solutions are $\underline{x = 2}$ and $\underline{x = {}^-4}$.

477. The equation is already in the proper form.

First, list the values for a, b, and c. $a = 2 \quad b = {}^-7 \quad c = {}^-30$

Substitute the values into the quadratic formula. $x = \frac{^-(^-7) \pm \sqrt{(^-7)^2 - 4(2)(^-30)}}{2(2)}$

Simplify the expression. $x = \frac{7 \pm \sqrt{49 - (^-240)}}{4} = \frac{7 \pm \sqrt{289}}{4} = \frac{7 \pm 17}{4}$

Find the two solutions for x by adding and then subtracting in the numerator.

$x = \frac{7 + 17}{4} = \frac{24}{4} = 6$ and

$x = \frac{7 - 17}{4} = \frac{^-10}{4} = {}^-2.5$

The two solutions are $\underline{x = 6}$ and $\underline{x = {}^-2.5}$.

478. The equation is already in the proper form.

First, list the values for a, b, and c. $a = 6 \quad b = 13 \quad c = {}^-28$

Substitute the values into the quadratic formula. $x = \frac{^-(13) \pm \sqrt{(13)^2 - 4(6)(^-28)}}{2(6)}$

Simplify the expression. $x = \frac{^-13 \pm \sqrt{169 + 672}}{12} = \frac{^-13 \pm \sqrt{841}}{12} = \frac{^-13 \pm 29}{12}$

Find the two solutions for x by adding and then subtracting in the numerator. $x = \frac{^-13 + 29}{12} = \frac{16}{12} = 1\frac{1}{3}$ and

$x = \frac{^-13 - 29}{12} = \frac{^-42}{12} = {}^-3\frac{1}{2}$

The two solutions are $x = 1\frac{1}{3}$ and $x = {}^-3\frac{1}{2}$.

479. The equation is already in the proper form.

First, list the values for a, b, and c. $a = 18 \quad b = 9 \quad c = 1$

Substitute the values into the quadratic formula. $x = \frac{^-(9) \pm \sqrt{(9)^2 - 4(18)(1)}}{2(18)}$

Simplify the expression. $x = \frac{^-9 \pm \sqrt{81 - 72}}{36} = \frac{^-9 \pm \sqrt{9}}{36} = \frac{^-9 \pm 3}{36}$

Find the two solutions for x by adding and then subtracting in the numerator. $x = \frac{^-9 + 3}{36} = \frac{^-6}{36} = \frac{^-1}{6}$ and

$x = \frac{^-12}{36} = \frac{^-1}{3}$

The two solutions are $x = \frac{^-1}{6}$ and $x = \frac{^-1}{3}$.

480. First, transform the equation into the proper form. Subtract the expression $(6 - 27x)$ from both sides of the equation. $36x^2 - 8x - (6 - 27x) = (6 - 27x) - (6 - 27x)$

Simplify both sides of the equation. $36x^2 - 8x - 6 + 27x = 0$

$36x^2 + 19x - 6 = 0$

Now, list the values of a, b, and c. $a = 36,\ b = 19,\ c = {}^-6$

Substitute the values into the quadratic formula. $x = \frac{^-19 \pm \sqrt{19^2 - 4(36)(^-6)}}{2(36)}$

Simplify the expression. $x = \frac{^-19 \pm \sqrt{1{,}225}}{72} = \frac{^-19 - 35}{72}$

Find the two solutions for x by adding and then subtracting in the numerator. $x = \frac{^-19 - 35}{72} = \frac{^-54}{72} = \frac{^-3}{4}$

$x = \frac{^-19 - 35}{72} = \frac{16}{72} = \frac{2}{9}$

The two solutions are $x = \frac{-3}{4}$.

481. First transform the equation into the proper form. Add ($^-12x - 32$) to both sides of the equation. $14x^2 - 12x - 32 = 12x + 32 - 12x - 32$

Combine like terms. $14x^2 - 12x - 32 = 0$

Now list the values for a, b, and c. $a = 14 \quad b = {}^-12 \quad c = {}^-32$

Substitute the values into the quadratic formula. $x = \frac{^-(^-12) \pm \sqrt{(^-12)^2 - 4(14)(^-32)}}{2(14)}$

Simplify the expression. $x = \frac{12 \pm \sqrt{144 + 1{,}792}}{28} = \frac{12 \pm \sqrt{1{,}936}}{28} = \frac{12 \pm 44}{28}$

Find the two solutions for x by adding and then subtracting in the numerator. $x = \frac{12 + 44}{28} = \frac{56}{28} = 2$ and

$x = \frac{12 - 44}{28} = \frac{^-32}{28} = {}^-1\frac{1}{7}$

The two solutions are $\underline{x = 2}$ and $\underline{x = {}^-1\frac{1}{7}}$.

Note that you could have divided both sides of the equation by 2 before listing your a, b, and c values. However, the solution would have been the same. Try it yourself.

482. The equation may not appear to be in proper form because there is no value for c. But you could write it as $4x^2 + 5x + 0 = 0$, and then your values would be $a = 4$, $b = 5$, and $c = 0$.

Substitute the values into the quadratic formula. $x = \frac{^-(5) \pm \sqrt{(5)^2 - 4(4)(0)}}{2(4)}$

Simplify the expression. $x = \frac{^-5 \pm \sqrt{25 - 0}}{8} = \frac{^-5 \pm 5}{8}$

Find the two solutions for x by adding and then subtracting in the numerator. $x = \frac{^-5 + 5}{8} = \frac{0}{8} = 0$ and

$x = \frac{^-5 - 5}{8} = \frac{^-10}{8} = {}^-1\frac{1}{4}$

The two solutions are $x = 0$ and $x = {}^-1\frac{1}{4}$.

483. Subtract 27 from both sides of the equation. $5x^2 - 27 = 27 - 27$ or $5x^2 - 27 = 0$

In this equation, there appears to be no coefficient for the x term unless you realize that $0x = 0$.

So you could write the equation in the proper form like this: $5x^2 + 0x - 27 = 0$

Now list the values for a, b, and c. $a = 5 \quad b = 0 \quad c = {}^-27$

Substitute the values into the quadratic formula. $x = \frac{^-(0) \pm \sqrt{(0)^2 - 4(5)(^-27)}}{2(5)}$

Simplify the expression.

$$x = \frac{^\pm\sqrt{4 \cdot 5 \cdot 27}}{10} = \frac{^\pm\sqrt{4 \cdot 9 \cdot 5 \cdot 3}}{10} = \frac{^\pm 2 \cdot 3\sqrt{5 \cdot 3}}{10} = \pm\frac{6\sqrt{15}}{10} = \pm\frac{3\sqrt{15}}{5}$$

The two solutions are $x = {}^{\pm}\frac{3}{5}\sqrt{15}$

484. Transform the equation into the desired form. Subtract $18x$ and add 17 to both sides.

$$5x^2 - 18x + 17 = 18x - 18x - 17 + 17$$

Combine like terms on both sides.

$$5x^2 - 18x + 17 = 0$$

Now list the values for a, b, and c.

$$a = 5 \quad b = {}^{-}18 \quad c = 17$$

Substitute the values into the quadratic formula.

$$x = \frac{{}^{-}({}^{-}18) \pm \sqrt{({}^{-}18)^2 - 4(5)(17)}}{2(5)}$$

Simplify the expression.

$$x = \frac{18 \pm \sqrt{324 - 340}}{10} = \frac{18 \pm \sqrt{{}^{-}16}}{10}$$

Since there is no real number equal to the square root of a negative number, there are no solutions for this equation.

485. Now, list the values of a, b, and c.

$$a = 9,\ b = {}^{-}30,\ c = 25$$

Substitute the values into the quadratic formula.

$$x = \frac{{}^{-}({}^{-}30) \pm \sqrt{({}^{-}30)^2 - 4(9)(25)}}{2(9)}$$

Simplify the expression.

$$x = \frac{30 \pm \sqrt{0}}{18} = \frac{30}{18} = \frac{5}{3}$$

So, there is only one solution to this equation, namely $x = \frac{5}{3}$.

486. First, transform the equation into the proper form. Add the expression $5x$ to both sides of the equation.

$${}^{-}12x^2 + 2 + 5x = {}^{-}5x + 5x$$

Simplify both sides of the equation.

$${}^{-}12x^2 + 2 + 5x = 0$$

$$12x - 5x - 2 = 0$$

Now, list the values of a, b, and c.

$$a = 12,\ b = {}^{-}5,\ c = {}^{-}2$$

Substitute the values into the quadratic formula.

$$x = \frac{5 \pm \sqrt{({}^{-}5) - 4(12)({}^{-}2)}}{2(12)}$$

Simplify the expression.

$$x = \frac{5 - \sqrt{121}}{24} = \frac{5 \pm \boxtimes\boxtimes}{24}$$

Find the two solutions for x by adding and then subtracting in the numerator.

$$x = \frac{5 + 11}{24} = \frac{16}{24} = \frac{2}{3}$$

$$x = \frac{5 - 11}{24} = \frac{{}^{-}6}{24} = {}^{-}\frac{1}{4}$$

The two solutions are $x = \frac{2}{3}$ and $x = {}^{-}\frac{2}{3}$.

487. First, multiply both sides of the equation by 2. $2x^2 = {}^-5x - 2$

Then add $5x + 2$ to both sides of the equation. $2x^2 + 5x + 2 = 0$

List the values of a, b, and c. $a = 2 \quad b = 5 \quad c = 2$

Substitute the values into the quadratic formula. $x = \frac{^-(5) \pm \sqrt{(5)^2 - 4(2)(2)}}{2(2)} = \frac{^-5 \pm \sqrt{25 - 16}}{2(2)} = \frac{^-5 \pm 3}{4}$

Find the two solutions for x by adding and then subtracting in the numerator. $x = \frac{^-5 + 3}{4} = \frac{^-2}{4} = \frac{^-1}{2}$ and $x = \frac{^-5 - 3}{4} = \frac{^-8}{4} = {}^-2$

The two solutions are $\underline{x = \frac{^-1}{2}}$ and $\underline{x = {}^-2}$.

488. Transform the equation by subtracting 5 from both sides. $x^2 + 8x - 5 = 0$

List the values of a, b, and c. $a = 1 \quad b = 8 \quad c = {}^-5$

Substitute the values into the quadratic formula. $x = \frac{^-8 \pm \sqrt{8^2 - 4(1)(^-5)}}{2(1)}$

Simplify the expression. $x = \frac{^-8 \pm \sqrt{64 + 20}}{2} = \frac{^-8 \pm \sqrt{84}}{2}$

$x = \frac{^-8 \pm \sqrt{4 \cdot 21}}{2} = \frac{^-8 \pm 2\sqrt{21}}{2} = {}^-4 \pm \sqrt{21}$

The solutions are $\underline{x = {}^-4 \pm \sqrt{21}}$.

489. First, transform the equation into the proper form. Subtract the expression $(1 - x)$ from both sides of the equation. $(1 - x) - (1 - x) = 56x^2 - (1 - x)$

Simplify both sides of the equation. $0 = 56x^2 - 1 + x$

$56x^2 + x - 1 = 0$

Now, list the values of a, b, and c. $a = 56, b = 1, c = {}^-1$

Substitute the values into the quadratic formula. $x = \frac{^-1 \pm \sqrt{1^2 - 4(56)(^-1)}}{2(56)}$

Simplify the expression. $x = \frac{^-1 \pm \sqrt{225}}{112} = \frac{^-1 \pm 15}{112}$

Find the two solutions for x by adding and then subtracting in the numerator. $x = \frac{^-1 - 15}{112} = \frac{^-16}{112} = \frac{^-1}{7}$

$x = \frac{^-1 + 15}{112} = \frac{14}{112} = \frac{1}{8}$

The two solutions are $\underline{x = \frac{^-1}{7}}$ and $\underline{x = \frac{1}{8}}$.

490. Transform the equation into proper form.
Use the distributive property of multiplication on the right side of the equation. $23x^2 = 16x - 2$
Subtract $16x$ and add 2 to both sides. $23x^2 - 16x + 2 = 0$
List the values of a, b, and c. $a = 23 \quad b = {}^-16 \quad c = 2$
Substitute the values into the quadratic formula. $x = \frac{-(-16) \pm \sqrt{(-16)^2 - 4(23)(2)}}{2(23)}$
Simplify the expression.

$$x = \frac{16 \pm \sqrt{256 - 184}}{2(23)} = \frac{16 \pm \sqrt{72}}{2(23)} = \frac{16 \pm \sqrt{4 \cdot 9 \cdot 2}}{2(23)} = \frac{2 \cdot 8 \pm 2 \cdot 3\sqrt{2}}{2(23)} = \frac{8 \pm 3\sqrt{2}}{23} = \frac{8}{23} \pm \frac{3}{23}\sqrt{2}$$

The solutions are $\underline{x = \frac{8}{23} \pm \frac{3}{23}\sqrt{2}}$.

491. The equation is already in the proper form.
List the values of a, b, and c. $a = 1 \quad b = 10 \quad c = 11$
Substitute the values into the quadratic formula. $x = \frac{-(10) \pm \sqrt{(10^2 - 4(1)(11)}}{2(1)}$
Simplify the expression.

$$x = \frac{{}^-10 \pm \sqrt{100 - 44}}{2} = \frac{{}^-10 \pm \sqrt{56}}{2} = \frac{{}^-10 \pm \sqrt{4 \cdot 14}}{2} = \frac{{}^-10 \pm 2\sqrt{14}}{2} = {}^-5 \pm \sqrt{14}$$

The two solutions are $\underline{x = {}^-5 + \sqrt{14}}$ and $\underline{x = {}^-5 - \sqrt{14}}$.

492. The Equation is already in the proper form.
List the values of a, b, and c. $a = 24 \quad b = 18 \quad c = {}^-6$
Substitute the values into the quadratic formula. $x = \frac{-(18) \pm \sqrt{(18)^2 - 4(24)(-6)}}{2(24)}$
Simplify the expression. $x = \frac{{}^-18 \pm \sqrt{324 + 576}}{48} = \frac{{}^-18 \pm \sqrt{900}}{48} = \frac{{}^-18 \pm 30}{48}$
Find the two solutions for x by adding and then subtracting in the numerator.

$x = \frac{{}^-18 + 30}{48} = \frac{12}{48} = \frac{1}{4}$ and
$x = \frac{{}^-18 - 30}{48} = \frac{{}^-48}{48} = {}^-1$

The two solutions are $\underline{x = \frac{1}{4}}$ and $\underline{x = {}^-1}$.

493. Transform the equation into the proper form. First use the distributive property of multiplication. $7x^2 = 4(3x) + 4(1) = 12x + 4$

Then subtract $(12x + 4)$ from both sides. $7x^2 - 12x - 4 = 0$

List the values of a, b, and c. $a = 7 \quad b = {}^{-}12 \quad c = {}^{-}4$

Substitute the values into the quadratic formula. $x = \frac{{}^{-}({}^{-}12) \pm \sqrt{({}^{-}12)^2 - 4(7)({}^{-}4)}}{2(7)}$

Simplify the expression. $x = \frac{12 \pm \sqrt{144 + 112}}{14} = \frac{12 \pm \sqrt{256}}{14} = \frac{12 \pm 16}{14}$

Find the two solutions for x by adding and then subtracting in the numerator. $x = \frac{12 + 16}{14} = \frac{28}{14} = 2$ and

$x = \frac{12 - 16}{14} = \frac{{}^{-}4}{14} = \frac{{}^{-}2}{7}$

The two solutions are $\underline{x = 2}$ and $\underline{x = \frac{{}^{-}2}{7}}$.

494. You could use the fractions as values for a and b, but it might be easier to first transform the equation by multiplying it by 12. $12(\frac{1}{3}x^2 + \frac{3}{4}x - 3 = 0)$

Using the distributive property, you get $12(\frac{1}{3}x^2) + 12(\frac{3}{4}x) - 12(3) = 12(0)$.

Simplify the terms. $4x^2 + 9x - 36 = 0$

List the values of a, b, and c. $a = 4 \quad b = 9 \quad c = {}^{-}36$

Substitute the values into the quadratic formula. $x = \frac{{}^{-}(9) \pm \sqrt{(9)^2 - 4(4)({}^{-}36)}}{2(4)}$

Simplify the expression. $x = \frac{{}^{-}9 \pm \sqrt{81 + 576}}{8} = \frac{{}^{-}9 \pm \sqrt{657}}{8}$

$= \frac{{}^{-}9 \pm \sqrt{9 \cdot 73}}{8} = \frac{{}^{-}9 \pm 3\sqrt{73}}{8}$

The two solutions for the variable x are $\underline{x = \frac{{}^{-}9 + 3\sqrt{73}}{8}}$ and $\underline{x = \frac{9 - 3\sqrt{73}}{8}}$.

495. First, transform the equation by multiplying both sides by 10 (to clear the decimals). Then, factor out a 5 from all terms, and divide both sides of the equation by 5.

$5x^2 + 15x - 20 = 0$

$5(x^2 + 3x - 4) = 0$

$x^2 + 3x - 4 = 0$

Now, list the values of a, b, and c. $a = 1, b = 3, c = {}^{-}4$

Substitute the values into the quadratic formula. $x = \frac{-3 \pm \sqrt{3^2 - 4(1)({}^{-}4)}}{2(1)}$

Simplify the expression. $x = \frac{-3 \pm \sqrt{25}}{2} = \frac{-3 \pm 5}{2}$

Find the two solutions for x by adding and then subtracting in the numerator.

$x = \frac{^-3-5}{2} = \frac{^-8}{2} = {}^-4$

The two solutions are $\underline{x = 1}$ and $\underline{x = -4}$.

496. List the values of a, b, and c. $a = 11 \quad b = {}^-4 \quad c = {}^-7$

Substitute the values into the quadratic formula.

$r = \frac{^-(^-4) \pm \sqrt{(^-4)^2 - 4(11)(^-7)}}{2(11)}$

Simplify the expression.

$r = \frac{4 \pm \sqrt{16 + 308}}{22} = \frac{4 \pm \sqrt{324}}{22} = \frac{4 \pm 18}{22}$

Find the two solutions for r by adding and then subtracting in the numerator.

$r = \frac{4 + 18}{22} = \frac{22}{22} = 1$ and

$r = \frac{4 - 18}{22} = \frac{^-14}{22} = {}^-0.64$

The two solutions are $\underline{r = 1}$ and $\underline{r \approx {}^-0.64}$.

497. List the values of a, b, and c. $a = 3 \quad b = 21 \quad c = {}^-8$

Substitute the values into the quadratic formula.

$m = \frac{^-(21) \pm \sqrt{(21)^2 - 4(3)(^-8)}}{2(3)}$

Simplify the expression.

$m = \frac{^-21 \pm \sqrt{441 + 96}}{6} = \frac{^-21 \pm \sqrt{537}}{6}$

The square root of 537 rounded to the nearest hundredth is 23.17. Substitute into the expression and simplify.

$m \approx \frac{^-21 + 23.17}{6} = \frac{2.17}{6} = 0.36$ and

$m \approx \frac{^-21 - 23.17}{6} = \frac{^-44.17}{6} = {}^-7.36$

The two solutions are $\underline{m \approx 0.36}$ and $\underline{m \approx {}^-7.36}$.

498. Transform the equation by subtracting $16y$ from each side and adding 5 to both sides.

$4y^2 - 16y + 5 = 16y - 5 - 16y + 5$

Combine like terms. $4y^2 - 16y + 5 = 0$

List the values of a, b, and c. $a = 4 \quad b = {}^-16 \quad c = 5$

Substitute the values into the quadratic formula.

$y = \frac{^-(^-16) \pm \sqrt{(^-16)^2 - 4(4)(5)}}{2(4)}$

Simplify the expression. $y = \frac{16 \pm \sqrt{256 - 80(1)}}{8} = \frac{16 \pm \sqrt{176}}{8} = \frac{16 \pm \sqrt{16 . 11}}{8} = \frac{16 \pm 4\sqrt{11}}{8}$

The square root of 11 rounded to the nearest hundredth in 3.32. Substitute into the expression and simplify.

$y = \frac{16 + 4(3.32)}{8} = \frac{29.28}{8} = 3.66$ and

$y = \frac{16 - 4(3.32)}{8} = \frac{2.72}{8} = 0.34$

The two solutions are $\underline{y \approx 3.66}$ and $\underline{y \approx 0.34}$.

499. List the values of a, b, and c. $a = 5 \quad b = 12 \quad c = {}^{-}1$

Substitute the values into the quadratic formula.

$$s = \frac{^{-}(12) \pm \sqrt{(12)^2 - 4(5)(^{-}1)}}{2(5)}$$

Simplify the expression.

$$s \approx \frac{^{-}12 \pm \sqrt{144 + 20}}{10} = \frac{^{-}12 \pm \sqrt{164}}{10}$$

The square root of 164 rounded to the nearest hundredth is 12.81. Substitute into the expression and simplify.

$$s \approx \frac{^{-}12 + 12.81}{10} = \frac{0.81}{10} = 0.081$$

$$\text{and } s \approx \frac{^{-}12 - 12.81}{10} = \frac{^{-}24.81}{10} = {}^{-}2.48$$

The two solutions for the variable s are $\underline{s \approx 0.08}$ and $\underline{s \approx {}^{-}2.48}$.

500. List the values of a, b, and c. $a = 4 \quad b = {}^{-}11 \quad c = 2$

Substitute the values into the quadratic formula.

$$z = \frac{^{-}(^{-}11) \pm \sqrt{(^{-}11)^2 - 4(4)(2)}}{2(4)}$$

Simplify the expression.

$$z = \frac{11 \pm \sqrt{121 - 32}}{8} = \frac{11 \pm \sqrt{89}}{8} \approx \frac{11 \pm 9.43}{8}$$

$$z = \frac{11 + 9.43}{8} = \frac{20.43}{8} = 2.55 \text{ and}$$

$$z = \frac{11 - 9.43}{8} = \frac{1.57}{8} = 0.20$$

The two solutions are $\underline{z \approx 2.55}$ and $\underline{z \approx 0.20}$.

501. List the values of a, b, and c. $a = 11 \quad b = {}^{-}32 \quad c = 10$

Substitute the values into the quadratic formula.

$$k = \frac{^{-}(^{-}32) \pm \sqrt{(^{-}32)^2 - 4(11)(10)}}{2(11)}$$

Simplify the expression.

$$k = \frac{32 \pm \sqrt{1{,}024 - 440}}{22} = \frac{32 \pm \sqrt{584}}{22} \approx \frac{32 \pm 24.17}{22}$$

$$k = \frac{32 + 24.17}{22} = \frac{56.17}{22} = 2.55 \text{ and}$$

$$k = \frac{32 - 24.17}{22} = \frac{7.8322}{22} = 0.36$$

The two solutions are $\underline{k \approx 2.55}$ and $\underline{k \approx 0.36}$.

Additional Online Practice

Whether you need help building basic skills or preparing for an exam, visit the LearningExpress Practice Center! On this site, you can access additional practice materials. Using the code below, you'll be able to log in and access additional online algebra practice. This online practice will also provide you with:

- **Immediate scoring**
- **Detailed answer explanations**
- **Personalized recommendations for further practice and study**

Log in to the LearningExpress Practice Center by using this URL: **www.learnatest.com/practice**

This is your Access Code: **8981**

Follow the steps online to redeem your access code. After you've used your access code to register with the site, you will be prompted to create a username and password. For easy reference, record them here:

Username: ____________________ Password: ______________________

With your username and password, you can log in and access the additional practice material. If you have any questions or problems, please contact LearningExpress customer service at 1-800-295-9556 ext.2, or e-mail us at customerservice@learningexpressllc.com.

Notes